# Was Hitler a Darwinian?

Charles Darwin (1809–1882). Albumen print, 1865–1866, by Ernest Edwards. (© National Portrait Gallery)

# Was Hitler a Darwinian?

## DISPUTED QUESTIONS IN THE HISTORY OF EVOLUTIONARY THEORY

Robert J. Richards

The University of Chicago Press CHICAGO & LONDON

**Robert J. Richards** is the Morris Fishbein Distinguished Service Professor of the History of Science and Medicine; professor in the Departments of History, Philosophy, Psychology, and in the Committee on Conceptual and Historical Studies of Science; and director of the Fishbein Center for the History of Science and Medicine, all at the University of Chicago.

The University of Chicago Press, Chicago 60637
The University of Chicago Press, Ltd., London

Printed in the United States of America

22 21 20 19 18 17 16 15 14 13    1 2 3 4 5

ISBN-13: 978-0-226-05876-4 (cloth)
ISBN-13: 978-0-226-05893-1 (paper)
ISBN-13: 978-0-226-05909-9 (e-book)
DOI: 10.7208/chicago/9780226059099.001.0001

Library of Congress Cataloging-in-Publication Data
Richards, Robert J. (Robert John), 1942–
Was Hitler a Darwinian? : disputed questions in the history of evolutionary theory / Robert J. Richards.
pages ; cm
Includes bibliographical references and index.
ISBN 978-0-226-05876-4 (cloth : alkaline paper) — ISBN 978-0-226-05893-1 (paperback : alkaline paper) — ISBN 978-0-226-05909-9 (e-book) 1. Evolution (Biology)—History. 2. Human evolution—Moral and ethical aspects. 3. Darwin, Charles, 1809–1882—Ethics. I. Title.
QH361.R534 2013
576.8'2—dc23
2013007451

♾ This paper meets the requirements of ANSI/NISO Z39.48-1992 (Permanence of Paper).

To Barbara

## CONTENTS

CHAPTER ONE

# Introduction

The past refuses to remain stable. This is due to its strange kind of existence—or rather, nonexistence—since past events no longer exist. Only the present exists, and, of course, the future is yet to be. So what kind of thing is the past? Is it merely what the actors of the time understood about their present—the objects, events, and individuals they experienced and thought about? Some historians maintain it is anachronistic to describe the past in terms other than those familiar to the persons of the period. This kind of practice would limit historians to what earlier individuals were aware of; scholars would thus be restricted to a narrow range of events and objects, only those falling under the actors' purview. Historians certainly want to discover how earlier individuals experienced their world. But, of course, some events of the past would not have been perceived correctly—at least, correct by our lights. Should we not try to get a handle also on those elusive events, articulating them, where appropriate, from the perspective of modern science?[1] Consider the Hippocratic physicians of the ancient period. They discriminated some three kinds of fever: those that spiked every other day, every third day, and every fourth day. Around those perceptions the ancient physicians draped an elaborate medical theory we no longer accept. While scrutinizing the features of that theory, might we be inclined to dismiss the observations of periodicity, thinking those early individuals to be under the sway of some numerical fantasy? We might

1. Someone like Jan Golinski seems to think we ought not apply contemporary science to help construe the past. See Golinski, *Making Natural Knowledge: Constructivism and the History of Science* (Cambridge: Cambridge University Press, 1998), 4: "Today's historians are more likely to set themselves the goal of understanding the past 'in its own terms' . . . rather than in the light of subsequent developments."

so reject their observations, if we did not know that malaria was endemic to Greece and that there are three strains of microorganism, each causing fevers to peak in the way the physicians described. It would be foolish not to use our contemporary knowledge as a way of determining what those past actors might actually have experienced, to show that the periodicity they ascribed to fevers was not completely tangled in the web of an antique imagination.

Then again, if we limited our descriptions to what individuals of the time might have recognized, we must ask: Which individuals? Were there millions of pasts but no single past? Presumably historians have the task of weaving together the experience of the world as lived by the pertinent players—and, at times, adjudicating: judging which historical individuals had a better grasp on the world and which deviated because of particular social, political, or religious convictions, or, in the case of naturalists, stumbled because of faulty instruments, poorly conceived experiments, or unconstrained imagination. Consider some of the organisms pictured in *Serpentum et draconum historiae libri duo* (1640), the posthumous work of the extraordinary, sixteenth-century naturalist Ulisse Aldrovandi (1522–1605). The first book provides illustrations of snakes of various sorts, some of monstrous birth (e.g., two-headed serpents); the text describes the earlier literature that had discussed a particular kind of serpent, the etymology of its name, its habits, medical uses, and so on. The second book describes, under similar headings, different kinds of dragon and includes illustrations of their many types (figs. 1.1 and 1.2). We know that dragons, those monsters of legend, don't exist, and it's foolish not to admit that. Our current knowledge allows us to explore the evidence that might have suggested the existence of dragons. It seems fairly certain that the creature in figure 1.1 is an African python—dragon-like enough. The small, two-footed dragon in figure 1.2 may well be based on the skeleton of a large, Indian fox-bat or the fossil remains of a pterodactyl. Aldrovandi had a collection of fossils in his cabinet of curiosity, so his belief in the mythical monsters he so naturalistically illustrated might well have had substantial grounds in direct observation—and the historian should lay out these possibilities, since they allow us to better understand the history of those times and the mentality of its individuals.[2] We are left only with traces of the past, but as Marc Bloch observed, "we

2. In her prizewinning book *Possessing Nature* (Berkeley: University of California Press, 1994), a scholarly account of the origins of natural history museums in early modern Italy, Paula Findlen begins her history with the tale of an omen that occurred on the occasion of the investiture of Pope Gregory XIII in 1572: "a fearsome dragon appeared in the countryside near Bologna" (17). She uses the appearance to introduce the collecting and descriptive activity of Ulisse Aldrovandi, relating how the event led to his production of the book on serpents and dragons and how this particular dragon provided a central attraction

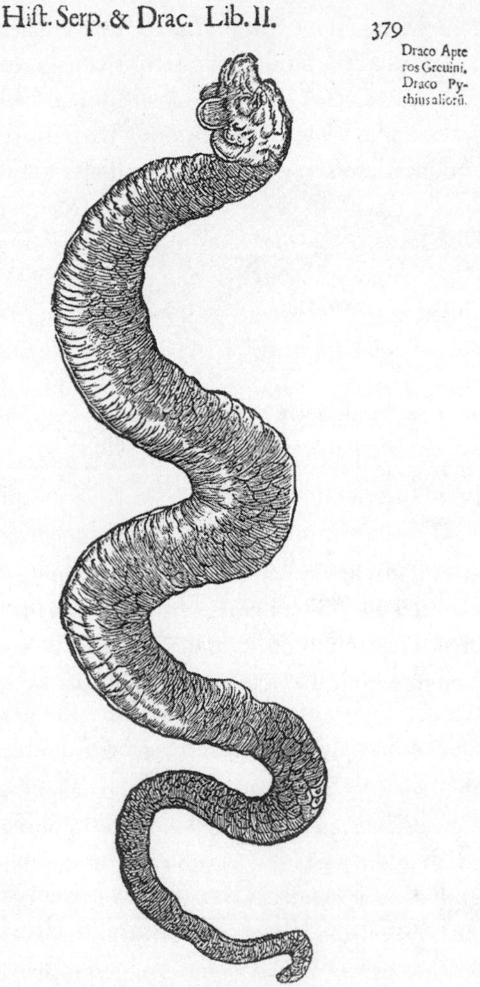


FIGURE 1.1 "Pythonic dragon." From Ulisse Aldrovandi, *Serpentum et draconum* (1640).

---

for his natural history collection. Nowhere in her recounting of this history (17–31) does she even hint that Aldrovandi brought into his collection something other than a real dragon. Her descriptions simply adopt Aldrovandi's point of view. Perhaps, though, she took it for granted that her readers knew dragons did not exist. Yet she makes no effort to suggest what he might actually have seen and what evidence he used to depict the many dragons displayed in his book. Her prose would lead the reader to believe that there really were dragons in those times—or to assume that Aldrovandi and his countrymen must have been suffering from mass hysteria.

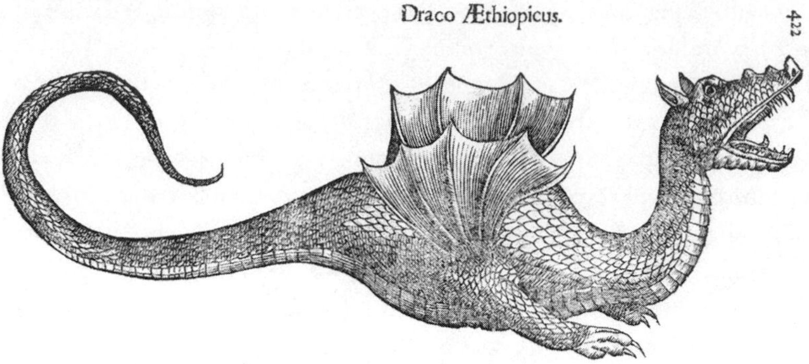


FIGURE 1.2 "Aethiopian dragon." From Ulisse Aldrovandi, *Serpentum et draconum* (1640).

are nevertheless successful in knowing far more of the past than the past itself had thought good to tell us."[3] The blanket proscription on the application of contemporary considerations to understand the past would only produce a crippled world, one in which the actors would hardly be recognizable as of our species.

The past comes into articulate existence only when historians gather evidence for events that no longer exist, when they construct those events in their accounts. But as new evidence becomes available and more reliable constructions are advanced, then the past, as we know it, changes, becomes other than it was. It's not just that interpretations of some perfectly articulated past time change. Rather, the only existence that the past has—as a creation of the historian—changes with new evidence or a richer understanding.

Though I am using the language of creation, bringing into existence a new past, I would not wish to deny an anchor for evidence and theory about the past. The situation is, I believe, much as the neo-Kantians have argued: there is a reality beyond the constructions of the human mind, but our only access to that reality is through the application of concepts to make events humanly tractable—the elusive shadow is made flesh through evidence and theory. The actors of the past deployed a web of concepts to grapple with nature and to bring it into living experience; historians, in their turn, also apply theory and

3. Marc Bloch, *The Historian's Craft*, trans. Peter Putnam (New York: Vintage Books, 1953), 63–64. Bloch was a founder of the *Annales* school of French historians, who used the various social sciences (e.g., demography, economics, etc.) to discover what the past might be forced to tell us.

evidence to bring not only the actors' experience into their narratives but also events beyond the actors' ken. Some evidence and concepts about past events will be better than others, will provide a more reliable guide in the construction of the past; and since the world—whether the social world or the natural—is not made of tapioca, the objects, the events, and the individuals with which historians deal will resist faulty constructions, which will then tumble to the ground when more adequate evidence and theory are advanced and slammed into place.

Though some historians may ignore these few historiographical principles, most simply follow them without much reflection on their epistemological import. And most historians of science, at any rate, conceive science as a special phenomenon, growing in accuracy and power as we approach the modern period. But what sets science history apart, distinct, at least, from the history of the arts, politics, and philosophy? And should we assume that it is so distinct? What is the evidence? The argument is simple and has been generally accepted, with some recent, notable demurs.

The historiographic assumption that there exists an extra-mental reality that provides stability to scientific endeavors is based on the inductive observation that scientific ideas (and historical constructions more generally), while changeable, are not radically so, as a naïve reader of Thomas Kuhn might come to believe. The Darwinian *revolution*, for instance, appears more like a Darwinian *evolution* when examined more closely. Some historians who would be loath to make this inductive conclusion to incremental advance contend there is nothing inevitable about scientific development. It has been argued that modern science, for example, need not have privileged experimental method, that the experimental method of contemporary science was merely the consequence of certain political and social processes in the seventeenth century. We could thus have had a modern science that used methods radically different from the experimental.[4]

Following the preceding line of thought, Gregory Radick, in an insightful, if objectionable, essay, considers whether Darwin's theory of natural selection was "inevitable."[5] Radick distinguishes two senses of this question: if Darwin had not read Malthus and had not held the social position he did, would he have come up with the idea of natural selection? And, if Darwin had not come

4. The locus classicus for this presumption is Steven Shapin and Simon Schaffer, *Leviathan and the Air-Pump* (Princeton: Princeton University Press, 1985).

5. Gregory Radick, "Is the Theory of Natural Selection Independent of Its History," in *The Cambridge Companion to Darwin*, 2nd ed., ed. Jonathan Hodge and Gregory Radick (Cambridge: Cambridge University Press, 2009), 147–72.

up with natural selection, would it nonetheless have been eventually discovered and consequently have shaped our contemporary science? Put another way, could we have a contemporary biology that remained ignorant of natural selection? Of course, Darwin might not have proposed the principle of natural selection; he could have died on the *Beagle* voyage or simply lost interest in natural history. So Radick answers no to the first question; Darwin might have followed his original plan and become a country parson, perhaps retaining his enthusiasm for earthworms but never having a thought about natural selection. Radick leaves the second question unanswered but hints at the answer he deems plausible. He maintains, similar to many social constructivist historians, that "people cannot be said to accept a theory *because* it is true. They may accept it because they believe the evidence shows the theory to be true, or because the theory is more parsimonious than its rivals, or because it fits well with prior beliefs and attitudes."[6] This view seems to imply that science floats on a cloud of beliefs not tethered to reality, a network of concepts that, in the case of modern biology, need not have included natural selection. Of course the historian may well ask—as in the case of Aldrovandi—why did the scientist *believe* the evidence and thus the theory to be true. It is perfectly arbitrary, after all, to allow unanchored beliefs to have causal potency and extra-mental nature to have none. Could it be that the evidence for a theory hooks onto those extra-mental structures, at least with elastic cords? Without that anchor, it seems perfectly inexplicable why scientific theories should work to the extent they do, or why historical development should appear more like an evolution than like a revolution. Even Kuhn came to recognize that evolution was a better model by which to explicate the history of science.[7] In their evolution, species respond to very different kinds of environment; mutatis mutandis, the evolution of scientific ideas would respond to the social, conceptual, and extra-mental environments. So according to this understanding, natural selection lay in wait, as it were, ready to be discovered. This consideration does not at all deny the potency of previous beliefs and the contributions of the social milieu. Rather it suggests that the source of scientific belief will usually be a matrix of causal vectors stemming from the social, the psychological, and the natural. In any particular instance the historian will be obliged to parcel out these vectors as best as he or she can.

6. Ibid., 162.

7. See Thomas Kuhn, *The Structure of Scientific Revolutions*, 2nd ed. (Chicago: University of Chicago Press, 1970), 172.

The essays that follow attempt to create a new past, pivoting around questions thought to have been settled in the history of biology. In constructing that past, I will not hesitate to deploy our contemporary understanding of certain features of a past of which the actors might have been only partially aware, or not aware at all. I will attempt, however, never to force a contemporary understanding on the actors; rather my effort will be to recognize the manner in which they understood events or applied certain concepts to nature, but also to bring causal analyses to explain why they likely construed events in the way they did, analyses that they themselves might not have been in a position to appreciate, indeed, accounts they might have rejected. So, for example, I will show that Darwin formulated his principle of divergence, a principle he thought as important as that of natural selection, in a way inconsistent with other aspects of his theory and that he was led down this shaded alley by his own practices as a pigeon fancier. I will point out that Darwin's principle, as he formulated it, and the auxiliary ideas associated with it, ill conform to our present knowledge of evolution. I will argue that Darwin's original principle of natural selection also had features that an older and less sanguine Darwin would likely have rejected—had he been fully cognizant of them. The essays in this volume, I have no doubt, will not be the last word on the questions pursued, and so they are offered as disputed questions in the history of evolutionary theory.

For contemporary ears, "disputed questions" sounds anomalous. Isn't it the answers to certain questions that are disputed? The phrase has a venerable history, characterizing as it does the mode of debate in medieval universities. In the *quaestiones disputatae* (investigations in dispute), a thesis would be posed by a master, and then two students, one denying the thesis (the *opponens*) and another affirming it (the *respondens*), would engage one another in vigorous debate. The master would sum up the merits of the arguments and make a decision as to the winner (the *determinatio*).[8] In the essays that follow, the role of the *respondens* will generally be played by those representing established scholarship on the theses considered. As *opponens*, I will try to show why closer analysis and evidence grounded in the texts under review will yield a determination rather different than that usually assumed—the results being, if I am successful, that a new past will swim into view.

Most canonical accounts take for granted that Darwin introduced blind mechanism into the explanation of biological phenomena, with the result that

8. See Brian Lawn, *The Rise and Decline of Scholastic "Quaestio Disputata"* (Leiden: Brill, 1993), 13–15.

nature gradually became drained of intelligence and moral value. Darwin, it is supposed, constructed an indifferent, materially neutral nature, one no longer posing as a surrogate for God and thus now become teleologically vacuous: man was dethroned from the center of the cosmos by Copernicus and now has slipped from the peak of a divinely constituted nature. Hasn't it been well established that Darwin's mechanism of natural selection has rendered human beings a lot less than the angels, indeed, no better than animals? As a consequence, can it be surprising that Hitler's extermination program adopted the idea of *struggle for existence* as a guiding principle? Hasn't it been made clear that Ernst Haeckel, Darwin's German disciple, conducted a fraudulent science, abetted the degradation of human beings, and contributed to the ideology of the Nazis? In the essays that follow, I will show that the appropriate answer to these questions is no.

The aforementioned are among the more contentious questions that the essays in this volume undertake to reexamine. Their resolution depends on more fundamental issues in the historical scholarship of evolutionary theory in the early period, also presumptively settled: the character of Darwin's chief principles of natural selection and divergence; the logical connection of his conception of natural selection to that of common descent; his dispute with Wallace over man's big brain; the role of language in human development; his relationship to Spencer; and the general problem of progress in evolution. In the case of Ernst Haeckel, the settled view has been that he committed fraud in the depiction of his thesis of recapitulation and that his artistic practice contributed to rendering his science decidedly subjective, whereas the more forward looking scientists of the period introduced modes of mechanical objectivity in their depictions of nature.[9] These too will be issues I reexamine.

Several of the essays in this volume will pursue the question of the moral character of evolutionary theory. I will attempt to show that Darwin did not regard the natural process of evolution as morally neutral. He wielded his device of natural selection in the *Origin of Species*, I will argue, to fix nature with a moral spine, a nature that, as depicted in the *Descent of Man*, has produced an animal that can make authentically moral choices. The evolutionary ethics that Darwin elaborated in the *Descent* captures what we intuitively understand of conscience and ethical behavior. This conception of Darwin's accomplishment runs against the grain of orthodox assumptions. Scholars such as Richard Dawkins, Michael Ghiselin, and Michael Ruse represent Darwinian man as always self-aggrandizing, always selfish in behavior—beneath the shell of

9. Lorraine Daston and Peter Galison, *Objectivity* (New York: Zone Books, 2007), 194–95.

other-regarding virtue, a core of original sin. Other critics, especially those of a politically or religiously conservative inclination, claim that this presumptive Darwinian construction of human nature was appropriated by Hitler with horrific consequence. I will dispute these conclusions and draw some unexpected, if limited, support from a recently published book by Peter Bowler, who in most respects adheres to an orthodox conception of Darwin's accomplishment.[10]

Bowler, like Radick, tries to imagine what the biological world would have been like had Darwin not lived. He recognizes that evolution as a branching phenomenon would have likely taken hold in the biological community of the mid-nineteenth century, but that Darwin's principle device of natural selection would not have emerged until quite a bit later, after the development of Mendelian genetics in the early twentieth century. Yet, bereft of Darwin, according to Bowler, the ethical outcome often attributed to him—the social Darwinism and eugenics that played through America, Britain, and especially Germany at the end of the nineteenth century—would nonetheless have occurred much in the way that it did. All of this is simply to say that Darwin's theory was not responsible for the malign social consequences often associated with his name. Bowler reaches this conclusion through the elaboration of a counterfactual history. I will advance toward a comparable conclusion but by a different route. We will otherwise differ markedly in our assessments of the theories of Darwin and his disciple Ernst Haeckel.

Several kinds of assumption have been responsible for the orthodox view of Darwinian evolutionary theory. One powerful set of such suppositions deploys contemporary science to articulate the past. This is not necessarily a distorting imposition. As I've already suggested, a judicious use of contemporary scientific theory can render a past visible, without which that past would remain shrouded in the earlier period's defective beliefs about nature. But there is also an injudicious use of scientific theory. It occurs because of a certain conviction about the nature of theory itself, a philosophical conceit often adopted by scientists, philosophers, and other scholars who have written about earlier evolutionary ideas. They assume that scientific theories are abstract, linguistic entities that can be instantiated at different times using even rather different terms that yet retain a common denotation. That view implies that the device of natural selection, for instance, would have an abstract formulation, which could be realized in nominally different terms at different periods but with

10. Peter Bowler, *Darwin Deleted: Imagining a World without Darwin* (Chicago: University of Chicago Press, 2013).

essentially the same meaning in, say, 1859 and 2013. Daniel Dennett, for instance, calls Darwin's device an "algorithm," thus fostering the presumption that what Darwin meant by natural selection is what we mean by it today.[11]

If we assume that theories are historical entities that develop over time, shaped by their environment—the environment of other theories and changing beliefs, as well as by natural events—then we are less likely to treat the concepts making up Darwin's original theory as essentially the same as those of today. We might then be more prepared to regard Darwin, though a harbinger of the modern age, yet a nineteenth-century thinker—a biologist who had not abandoned teleological ideas but conceived nature as having the goal of producing human beings. Well, conclusions of this kind require considerable evidence and argument, which the essays in this volume will lay out.

To regard scientific theories as historical entities has another consequence: it focuses our attention on the actual words of a text. Darwin's manuscripts and books are rife with the use of metaphors and tropes. We could regard these as merely rhetorical flourishes, not to be taken seriously, light camouflage disguising the real, universal logic of his theory. But once we understand the history of these metaphors in the context of the development of Darwin's ideas, we will see how deeply embedded they are in his understanding of nature and how they cannot, as George Levine has made clear, be "skimmed off" to expose the real science.[12] These tropes are frequently formulated to capture features of nature implicit in Darwin's conception, and in that respect the metaphors do real work in the theory. They reveal the logical structure actually undergirding Darwin's books, not a logic completely isomorphic with that of modern theory. An index of the power of these metaphors is exemplified in Darwin's only partial acceptance of Wallace's advice about the trope of "natural selection" itself. As I will describe more thoroughly in chapter 2, Wallace suggested that Darwin replace the phrase with Herbert Spencer's "survival of the fittest."[13] Wallace thought "natural selection" implied an intentional activity on the part of nature. Darwin demurred. He did introduce Spencer's phrase in the fifth

11. Daniel Dennett, *Darwin's Dangerous Idea: Evolution and the Meaning of Life* (New York: Simon and Schuster, 1995), 48–60. In 2009, at a symposium sponsored by the National Academy of Sciences, Dan and I both gave talks on Darwin's accomplishment. On the bus carrying the participants back from the meeting, we sat next to each other. As our dispute became louder, Dan capped it with: "I don't give a God-damn what Darwin said; it's not true."

12. See George Levine's sensitive study of Darwin's language in Levine, *Darwin the Writer* (New York: Oxford University Press, 2012).

13. Alfred Russel Wallace to Charles Darwin (2 July 1866), in *The Correspondence of Charles Darwin*, ed. Frederick Burkhardt et al., 19 vols. to date (Cambridge: Cambridge University Press, 1985–), 14:227–29.

edition of the *Origin* but also kept his original formulation, since it captured something he wanted to retain, though he could not explicitly articulate what that something was. As I will argue in the next two chapters, the phrase captured more than Darwin, later in his career, would care to emphasize. Yet the phrase and its metaphorical implications were so tightly woven into the fabric of his theory that it simply could not be cut out without the whole thing unraveling.

In the next two chapters of this volume, I consider the two principles that Darwin regarded as key to understanding his theory: natural selection and divergence. As I've suggested, my construction of these principles, as they operate in the *Origin of Species*, will differ from the orthodox and traditional conception of Darwin's accomplishment. Chapter 4 brings my deviating interpretation to the *Descent of Man*. In this chapter, in addition to explicating Darwin's proposal for an evolutionary ethics, I will also show the relevance of that proposal for our contemporary assessment of the character of moral judgment. Chapter 5 contrasts Darwin's view of evolution with that of his sometime rival Herbert Spencer. I believe their respective theories are more complementary than usually supposed. Chapters 6 and 7 concern Darwin's great disciple in Germany, Ernst Haeckel. Haeckel brought an artist's sensibilities to the science of biology, a sensibility that may have led him to a portrayal of nature that reached beyond the ken of his more empiricist opponents. I believe that in our time a great injustice has been done to the reputation of this extraordinary thinker. Haeckel provides a cautionary case against the imposition of contemporary science on the past, in so far as that imposition has fallen prey to bad history, bad logic, and bad judgment. Chapter 8 shows how Haeckel's friend, the linguist August Schleicher, brought the science of language to help secure transmutational theory and also to furnish a distinctive entrée into the human mind; Schleicher's theory of language development led to Darwin's and Haeckel's reciprocal notions of human mental development. Chapter 9, the final chapter, which gives this collection its title, provides a synthesis of the previous chapters, but more directly seeks to show that both contemporary scholars and those of a religious disposition have made ideologically driven connections between Nazi biology and Darwinian theory.

No theory has so saturated the scientific mind with productive possibilities as Darwin's has, yet outside the scientific community no theory has evoked such deeply felt negative reactions. Even some scientists and philosophers who embrace the theory in the domain of biology refuse to go the whole orang with Darwin, stopping short of human mind and morals. In the essays that follow, I have tried to show that Darwinian theory, at least in its original composition, does not have the soul-deadening or dangerous consequences that many have supposed, on both the conservative and liberal sides. Conservatives have to face the theory squarely, recognizing that it is supported by unparalleled evidence and that it hardly denigrates man but elevates nature. Liberals also need to be more relaxed: its application to human beings, even construing the mind's highest powers, does not lead to either racism or moral delinquency. Moreover, the liberal has no choice: human beings have not escaped the pervading powers of nature. But as I hope to show in these essays, nature may not be, at least for the flexible of mind, the sterile, irrational process that even some neo-Darwinists have supposed it to be.

All of the essays that follow have been published within the last ten years or so, except "Was Hitler a Darwinian?" which appears for the first time. I have thoroughly revised the essays, frequently adding material or combining two previously published essays into one. I have tried to reduce redundancies as much as possible, while keeping each of the essays independently intelligible; inevitably though, the essays reiterate some arguments. The translations, unless otherwise noted, are my own. I do not doubt that many of the questions at issue will remain unsettled. But I am warmed by the maxim derived from another intellectual activity: "Faint heart never filled a flush."

CHAPTER TWO

# Darwin's Theory of Natural Selection and Its Moral Purpose

Thomas Henry Huxley (1825–1895; fig. 2.1) recalled that after he had read Darwin's *Origin of Species* (1859), he exclaimed to himself: "How extremely stupid not to have thought of that!"[1] It is a famous but puzzling remark. In his contribution to Francis Darwin's *Life and Letters of Charles Darwin*, Huxley rehearsed the history of his engagement with the idea of transmutation of species. He mentioned the views of Robert Grant (1793–1874), an advocate of Lamarck (1744–1829), and Robert Chambers (1802–1871), who anonymously published *Vestiges of the Natural History of Creation* (1844), which advanced a crude idea of transmutation. He also recounted his rejection of Louis Agassiz's (1807–1873) belief that species were progressively replaced by the divine hand. He neglected altogether his friend Herbert Spencer's (1820–1903) early Lamarckian ideas about species development, which were also part of the long history of his encounters with the theory of descent. None of these sources moved him to adopt any version of the transmutation hypothesis.

Huxley was clear about what finally led him to abandon his long-standing belief in species stability: "The facts of variability, of the struggle for existence, of adaptation to conditions, were notorious enough; but none of us had suspected that the road to the heart of the species problem lay through them, until Darwin and Wallace dispelled the darkness, and the beacon-fire of the 'Origin' guided the benighted."[2]

1. Thomas Henry Huxley, "The Reception of the *Origin of Species*," in *Life and Letters of Thomas Henry Huxley*, 2 vols., ed. Leonard Huxley (New York: D. Appleton, 1900), 1:183.

2. Ibid., 183.

FIGURE 2.1 Thomas Henry Huxley (1826–1895), in 1857. Photo from Leonard Huxley, *Life and Letters of Thomas H. Huxley.*

The elements that Huxley indicated—variability, struggle for existence, adaptation—form core features of Darwin's conception of natural selection. Thus what Huxley admonished himself for not immediately comprehending was not the fact of species change but the cause of that change. Huxley's exclamation suggests—and it has usually been interpreted to affirm—that the idea of natural selection was really quite simple and that when the few elements composing it were held before the mind's eye, the principle and its significance would flash out. The elements, it is supposed, fall together in this way: species members vary in their heritable traits from each other; more individuals are produced than the resources of the environment can sustain; those that by chance have traits that better fit them to their circumstances than others of their kind will more likely survive to pass on those traits to offspring; consequently, the structural character of the species will continue to alter over generations until individuals appear specifically different from their ancestors.[3]

Yet if the idea of natural selection were as simple and fundamental as Huxley suggested and as countless scholars have maintained, why did it take so long for the theory to be published after Darwin supposedly discovered it? And why did it then require a very long book to make its truth obvious? In this chapter, I will try to answer these questions. I will do so by showing that the principle of natural selection is not simple but complex and that it only gradually took shape in Darwin's mind. In what follows, I will refer to the "principle" or "device" of natural selection, never the "mechanism" of selection. It is widely assumed that a singular feature of Darwin's accomplishment was that he introduced mechanism into nature. In a typical fashion, Richard Lewontin, Steven Rose, and Leon Kamin so render the identifying construct of Darwin's science: "Natural selection theory and physiological reductionism were explosive and powerful enough statements of a research program to occasion the replacement of one ideology—of God—by another: a mechanical, materialist science."[4] Though the phrase "mechanism of natural selection" comes trippingly to our tongues, it never came to Darwin's. Yet even when the focus is directly on the historical Darwin, scholars almost reflexively use this locution, thereby making the slide to a metaphysical conviction much easier.[5] I will not hesitate to use the term "evolution" to describe the idea of species descent with modification. Somehow the notion has gained currency that Darwin avoided the term because it suggested progressive development.[6] This assumption has

3. Waters succinctly provides the standard account in three principles: (1) variations appear in organisms without preadaptation to the environment; (2) some variations by chance work in the environment and give bearers an advantage over those lacking the traits; and (3) such variations are usually transmitted to progeny. Waters offers a comparably succinct and generally orthodox account of Darwin's entire argument. See C. Kenneth Waters, "The Arguments in the *Origin of Species*," in *Cambridge Companion to Darwin*, 2nd ed., ed. Jonathan Hodge and Gregory Radick (Cambridge: Cambridge University Press, 2009), 120–43; his distillation of natural selection is on p. 128. I do not doubt that these principles capture essential features of Darwin's idea, but I argue that this abstract formulation misses much else in his conception of natural selection.

4. R. C. Lewontin, Steven Rose, and Leon Kamin, *Not in Our Genes* (New York: Pantheon, 1984), 51.

5. For example, see Michael Ruse, *Darwinism and Its Discontents* (Cambridge: Cambridge University Press, 2008), in which the linguistic reflex "mechanism of natural selection" is given ample play—some nineteen times in a moderately sized book.

6. Richard Lewontin, for one, claims that Darwin did not use the term "evolution" because it suggested a progressive development of organism, whereas his theory rejected a progressivist view. See my exchange with Lewontin, "Darwin and Progress," *New York Review of Books* 52, no. 20 (15 December 2005), letters.

no warrant for two reasons. First, the term is obviously present, in its participial form, as the very last word in the *Origin*, as well as being freely used as a noun in the last edition of the *Origin* (1872), in the *Variation of Animals and Plants under Domestication* (1868), and throughout the *Descent of Man* (1871) and the *Expression of the Emotions in Man and Animals* (1872). But the second reason for rejecting the assumption is that Darwin's theory is, indeed, progressivist, and his device of natural selection was designed to produce evolutionary progress.

Scholars have supposed that a red thread runs through a progressivist interpretation of nature, leading to the assumption that human beings are the goal of nature's strivings, an assumption they believe to be a remnant of antique theology and quite antithetic to Darwin's intentions. I rather believe that Darwin constructed his theory precisely with this teleological trajectory in mind. In this chapter, I argue that Darwin cast natural selection as the device by which, as he put it, "the most exalted object we are capable of conceiving" has been achieved: man as a moral creature. To trace the thread and determine its endpoint, one must start at the beginning of Darwin's theorizing.

## DARWIN'S EARLY EFFORTS TO EXPLAIN SPECIES TRANSFORMATION, 1837–1838

Shortly after he returned from his voyage on H.M.S. *Beagle* (1831–36; fig. 2.2), Darwin began seriously to entertain the hypothesis of species change over time. He had been introduced to the idea, when a teenager, through reading his grandfather Erasmus Darwin's *Zoonomia* (1794–96), which included speculations about species development; while at medical school in Edinburgh (1825–27), he studied Lamarck's *Système des animaux sans vertèbres* (1801) under the tutelage of Robert Grant, a convinced evolutionist. On the voyage, he packed into his cabin Lamarck's *Histoire naturelle des animaux sans vertèbres* (1815–22), in which the idea of evolutionary change was prominent. He got another large dose of the Frenchman's ideas during his time off the coast of South America, where he received by merchant ship the second volume of Charles Lyell's *Principles of Geology* (1831–33), which contained a searching discussion and negative critique of the fanciful supposition of an "evolution of one species out of another."[7] Undoubtedly the rejection of Lamarck by Lyell and most British naturalists gave Darwin pause; after his return to England,

7. Charles Lyell, *Principles of Geology*, 3 vols. (1830–33; repr., Chicago: University of Chicago Press, 1987), 2:60.

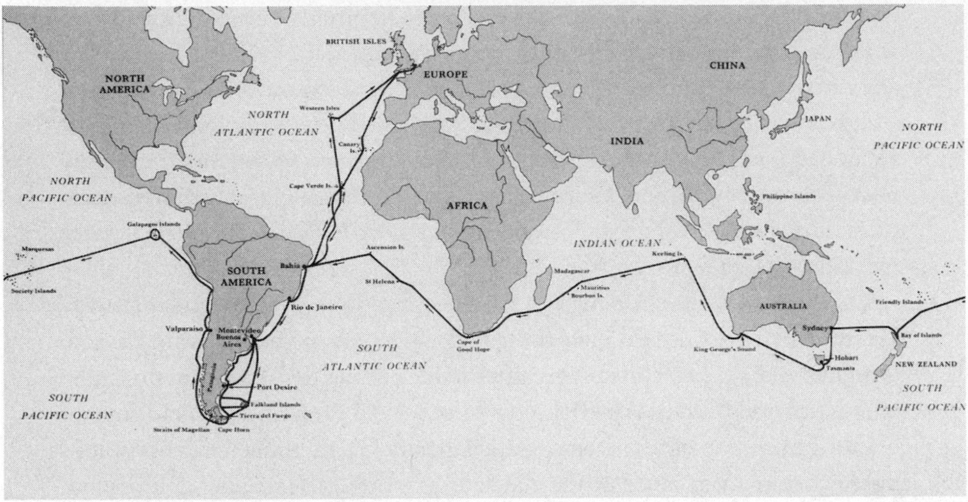


FIGURE 2.2 Voyage of H.M.S. *Beagle*. Departed Plymouth, December 1831; returned Falmouth, October 1836.

however, while sorting and cataloging his specimens from the Galapagos, he came to understand that his materials supplied compelling evidence for the suspect theory. Three groups of mockingbirds, which he had thought merely varieties of the mainland species, were identified by John Gould (1804–1881), chief ornithologist of the British Museum, as distinct species.[8] The revelation tripped a mind at the ready.

In his various early notebooks (January 1837–June 1838), Darwin began to work out different possibilities to explain species change.[9] Initially, he supposed that a species might be "created for a definite time," so that when its span of years was exhausted, it went extinct and another, affiliated species took its place.[10] He rather quickly abandoned the idea of species senescence and began to think in terms of Lamarck's notion of the direct effects of the environment, especially the possible impact of the imponderable fluids of heat and

8. Sulloway has persuasively argued that it was Gould's identification that convinced Darwin of the transmutational theory. See Frank Sulloway, "Darwin's Conversion: The *Beagle* Voyage and Its Aftermath," *Journal of the History of Biology* 15 (1982): 327–98.

9. Robert J. Richards, *Darwin and the Emergence of Evolutionary Theories of Mind and Behavior* (Chicago: University of Chicago Press, 1987), 85–98.

10. Charles Darwin, *Red Notebook* (MS p. 129), in *Charles Darwin's Notebooks, 1836–1844*, ed. Paul Barrett et al. (Ithaca: Cornell University Press, 1987), 62.

electricity.[11] If the device of environmental impact were to meet what seemed to be the empirical requirement—as evidenced by the pattern of fossil deposits, going from simple shells at the deepest levels to complex vertebrate remains at higher levels—then it had to produce progressive development. If species resembled ideas, then progressive change would seem to be a natural result, or so Darwin speculated: "Each species changes. Does it progress. Man gains ideas. The simplest cannot help.—becoming more complicated; & if we look to first origin there must be progress."[12] Being the conservative thinker that he was, Darwin retained in the *Origin* the idea that some species, under special conditions, might be transformed through direct environmental impact; at a deeper level in the book, his progressivist conviction, persisting from this early period, provided his theory a definite vector for the evolution of organisms.

Darwin seems to have soon recognized that the direct influence of surroundings on an organism could not account for its more complex adaptations, and so he began constructing another causal device. He had been stimulated by an essay of Frédéric Cuvier's (1773–1838), the great Georges Cuvier's (1769–1832) younger brother; the essay suggested that animals might acquire heritable traits through exercise in response to particular circumstances. Darwin quickly concluded that "all structures either direct effect of habit, or hereditary ‹& combined› effect of habit."[13] He thus assumed that new habits, if practiced by a population over long periods of time, would turn into instincts, and that these latter would eventually modify anatomical structures and so would alter species. Use-inheritance was, of course, a principal mode of species transformation for Lamarck.

In developing his own theory of use-inheritance, Darwin carefully distinguished his ideas from those of his discredited predecessor—or at least he was persuaded that their respective ideas were quite different. He attempted to distance himself from the French naturalist by proposing that habits introduced into a population would first gradually become instinctual before they altered anatomy. And instincts—innate patterns of behavior—would be expressed automatically, without the intervention of conscious willpower, the presumptive Lamarckian mode.[14] By early summer of 1838, Darwin thus had two devices by which to explain descent of species with modification: the direct effects of the environment and his habit-instinct device.

11. Charles Darwin, *Notebook B* (MS pp. 17–20), in *Charles Darwin's Notebooks*, 175.

12. Ibid. (MS p. 18).

13. Charles Darwin, *Notebook C* (MS p. 63), in *Charles Darwin's Notebooks*, 259. (The editors of the *Notebooks* use double wedges to indicate insertions by Darwin.)

14. Ibid. (MS p. 171), 292.

## ELEMENTS OF THE THEORY OF NATURAL SELECTION

At the end of September 1838, Darwin paged through Thomas Malthus's (1766–1834; fig. 2.3) *Essay on the Principle of Population* (6th ed., 1826). As he later recalled in his *Autobiography*, this happy event changed everything for his developing conceptions:

FIGURE 2.3 Thomas Robert Malthus (1766–1834). Mezzotint. (© National Portrait Gallery)

> I soon perceived that selection was the keystone of man's success in making useful races of animals and plants. But how selection could be applied to organisms living in a state of nature remained for some time a mystery to me. In October 1838, that is, fifteen months after I had begun my systematic enquiry, I happened to read for amusement Malthus on Population, and being well prepared to appreciate the struggle for existence which everywhere goes on from long-continued observation of the habits of animals and plants, it at once struck me that under these circumstances favourable variations would tend to be preserved, and unfavourable ones to be destroyed. The result of this would be the formation of new species. Here, then, I had at last got a theory by which to work.[15]

Darwin's description supplies the classic account of his discovery, and it does capture a moment of that discovery, though not the complete character or full scope of his mature conception. The account in the *Autobiography* needs to be placed against the notebooks, essays, and various editions of the *Origin* and the *Descent of Man*. These comparisons reveal *many* moments of discovery, and a gradual development of his theory of natural selection from 1838 through the next several decades.

In the *Autobiography*, Darwin mentioned two considerations that had readied him to detect in Malthus a new possibility for the explanation of species development: the power of artificial selection and the role of struggle. Lamarck had suggested domestic breeding as the model for what occurred in nature. Undeterred by Lyell's objection that domestic animals and plants were specially created for man, Darwin sought guidance for determining how selection might operate in nature from breeders' manuals, such as those by John Sebright (1767–1846) and John Wilkinson (1797–1875).[16] This literature brought him to understand more clearly the power of domestic "selection"

15. Charles Darwin, *The Autobiography of Charles Darwin*, ed. Nora Barlow (New York: Norton, 1969), 119–20. Darwin's *Autobiography* puts the Malthusian moment in October 1838, but his notebooks testify that the inspiration came a bit earlier, at the end of September of that year.

16. Lyell, *Principles of Geology*, 2:41. John Sebright, *The Art of Improving the Breeds of Domestic Animals* (London: Howlett and Brimmer, 1809); John Wilkinson, "Remarks on the Improvement of Cattle, etc. in a Letter to Sir John Sanders Sebright, Bart. M.P.," Nottingham, 1820. Ruse shows how these breeders contributed to Darwin's understanding of the nature of artificial selection. See Michael Ruse, "Charles Darwin and Artificial Selection," *Journal of the History of Ideas* 36 (1975): 339–50. For an expansive review of Darwin's notions about artificial selection, see Bert Theunissen, "Darwin and His Pigeons, the Analogy between Artificial Selection and Natural Selection Revisited," *Journal of the History of Biology* 45 (2012):179–212.

(Sebright's term), but he remained puzzled, as his *Autobiography* suggests, about what might play the role of the natural selector or "picker." In midsummer of 1838, he observed: "The Varieties of the domesticated animals must be most complicated, because they are partly local & then the local ones are taken to fresh country & breed confined, to certain best individuals.—scarcely any breed but what some individuals are picked out.—in a really natural breed, not one is picked out."[17] This passage illustrates Darwin's perplexity: How could selection occur in nature when no agent was picking the few "best individuals" to breed?

In the *Autobiography*, Darwin indicated that the second idea that prepared him to divine the significance of Malthus's *Essay* was that of the struggle for existence. Lyell, in the *Principles of Geology*, had mentioned the observation of Augustin de Candolle (1778–1841) that all the plants of a country "are at war with one another." This kind of struggle, Lyell believed, would be the cause of "mortality" of species, of which fossils gave abundant evidence.[18] In his own reading of Lyell, Darwin took to heart the implied admonition to "study the wars of organic beings."[19]

These antecedent notions gleaned from Lamarck, Lyell, and the breeders led Darwin to the brink of a stable conception. In spring of 1837, for instance, he considered how a multitude of varieties might yield creatures better adapted to circumstances: "whether every animal produces in course of ages ten thousand varieties, (influenced itself perhaps by circumstances) & those alone preserved which are well adapted."[20] Here—eighteen months before he read Malthus—Darwin mentioned in passing a central element of his principle of natural selection without, apparently, detecting its significance. And a year later, something like both natural and sexual selection spilled onto the pages of his *Notebook C*: "Whether species may not be made by a little more vigour being given to the chance offspring who have any slight peculiarity of structure. «(hence seals take victorious seals, hence deer victorious deer, hence males armed & pugnacious all orders; cocks all war-like)»."[21] It is fair to say, nonetheless, that the foundations for Darwin's device of natural selection were laid on

17. Charles Darwin, *Notebook D* (MS p. 20), in *Charles Darwin's Notebooks*, 337.

18. Lyell, *Principles of Geology*, 2:131, 130.

19. Darwin, *Notebook C* (MS p. 73), 262.

20. Darwin, *Notebook B* (MS p. 90), 193.

21. Darwin, *Notebook C* (MS p. 61), 258. This entry is likely a gloss on Sebright, *Art of Improving the Breeds*, 15–16.

the ground of Malthus's *Essay*. His reading of that book caused those earlier presentiments to settle into a firm platform for further development.

## THE MALTHUS EPISODE

Malthus had composed his book to investigate two questions: What has kept humankind from steadily advancing in happiness? And, can the impediments to happiness be removed? Famously, he argued that the chief barrier to the progress of civil society was that population increase would always outstrip the growth in the food supply, thus causing periodic misery and famine. What caught Darwin's eye in the opening sections of Malthus's *Essay*, as suggested by scorings in his copy of the book, was the notion of population pressure through geometric increase:

> In the northern states of America, where the means of subsistence have been more ample . . . the population has been found to double itself, for above a century and half successively, in less than twenty-five years. . . . It may safely be pronounced, therefore, that population, when unchecked, goes on doubling itself every twenty-five years, or increases in a geometrical ratio. . . . But the food to support the increase from the greater number will by no means be obtained with the same facility. Man is necessarily confined in room.[22]

Darwin found in those passages from Malthus a propulsive force that had two effects: it would severely restrict reproduction by reason of the better adapted pushing out the weaker and thus depriving them of resources, and consequently it would sort out, or transform, the population. On 28 September 1838, Darwin phrased it this way in his *Notebook D*:

> Even the energetic language of ‹Malthus› «Decandoelle» does not convey the warring of the species as inference from Malthus . . . population in increase at geometrical ratio in FAR SHORTER time than 25 years—yet until the one sentence of Malthus no one clearly perceived the great check amongst men. . . . One may say there is a force like a hundred thousand wedges trying force ‹into› every kind of adapted structure into the gaps ‹of› in the oeconomy of Nature, or rather forming gaps by thrusting out

22. Thomas R. Malthus, *An Essay on the Principle of Population*, 6th ed., 2 vols. (London: Murray, 1826), 1:5.

weaker ones. «The final cause of all this wedging, must be to sort out proper structure & adapt it to change».[23]

All the "wedging" caused by population pressure would have the effect, according to Darwin, of filtering out all but the most fit organisms and thus adapting them (actually, leaving them preadapted) to their circumstances. One should note, however, that Darwin does not emphasize the negative feature of this process, namely, the death of vast numbers of the population for lack of resources; rather he looks to the positive effect of sorting out and adapting the population. In the gradual construction of his theory, he constantly stressed the positive over the negative. He turned away from death.

Though natural selection is the linchpin of Darwin's theory of evolution, his notebooks indicate only the slow emergence of its ramifying features. He reflected on his burgeoning notions through the first week of October 1838 but then turned to other matters. Through the next few months, here and there, the implications became more prominent in his thought. In early December, for instance, he explicitly drew for the first time the analogy between natural selection and domestic selection: "It is a beautiful part of my theory, that «domesticated» races . . . are made by percisely [*sic*] same means as species."[24] But the most interesting reflections, which belie the standard assumptions about Darwin's theory, were directed to the final cause or purpose of evolution. This teleological framework would help organize several other elements constituting his developing notion.

## THE PURPOSE OF PROGRESSIVE EVOLUTION: THE MORAL ANIMAL

The *Origin of Species* concludes with a great peroration that Darwin had honed over several decades: "Thus, from the war of nature, from famine and death, the most exalted object which we are capable of conceiving, namely, the production of the higher animals directly follows. There is grandeur in this view of life, with its several powers, having been originally breathed into a few forms or into one; and that, whilst this planet has gone cycling on according to the fixed law of gravity, from so simple a beginning endless forms most beautiful and most wonderful have been, and are being, evolved."[25]

23. Darwin, *Notebook D* (MS p. 135e), 375–76.
24. Charles Darwin, *Notebook E* (MS pp. 71, 63), in *Charles Darwin's Notebooks*, 416, 414.
25. Charles Darwin, *On the Origin of Species* (London: Murray, 1859), 490.

In this lyrical conclusion, Darwin asserted a long-standing and permanent conviction, namely, that the "object," or purpose, of the "war of nature" is "the production of the higher animals." And the unspoken, but clearly intended, higher animals were human beings with their moral sentiments. Darwin imbedded his developing theory of natural selection in a decidedly progressivist and teleological framework, a framework quite obvious when one examines the initial construction of his theory.

Several passages from his early notebooks indicate Darwin's teleological perspective on the operations of natural selection in nature. As a coda to his reading of Malthus in late September 1838, he added this characteristic teleological mode of consideration: "The final cause of all this wedging, must be to sort out proper structure & adapt it to change—to do that, for form, which Malthus shows, is the final effect (by means of volition) of this populousness on the energy of Man."[26]

Darwin here construed the purpose of population pressure as the adaptation of organic form to changing circumstances. Thus at the very birth of the idea of natural selection, Darwin conceived the process as comparable to what happened when energetic colonists moved into new territories and intentionally drove out indigenous peoples.[27]

Darwin's use of the language of final causes might be thought only a *façon de parler*, something the careful historian need not take seriously. After all, many scholars have credited Darwin precisely with the abolition of final causes from nature. But we must bear in mind that Darwin, this herald of modern biology, was yet a nineteenth-century thinker. His conceptions were wrought in terms available to his time and circumstances. And he frequently enough deployed final causes as part of the explanation of natural processes. For example, when considering Lyell's descriptions of the virtually limitless geological periods before the appearance of man, Darwin recast them into a teleological account: "Progressive development gives final cause of enormous periods anterior to Man."[28] In other words, the purpose of the vast extents of time prior to the appearance of human beings was for the gradual progressive development of the necessary antecedent conditions. Or take a more salient example that appears in Darwin's notebook a month after his reading of Malthus. In an entry at the end of October 1838, Darwin considers how his theory could explain

26. Darwin, *Notebook D* (MS p. 135e), 375–76.

27. Malthus had argued that when populations grow large, the energetic offspring are urged to seek new territories, even those already settled by native societies. See Malthus, *Essay*, 1:94–95.

28. Darwin, *Notebook B* (MS p. 49), 182.

a puzzle that is still of interest—why sexual generation evolved instead of nature remaining satisfied with asexual modes of reproduction: "My theory gives great final cause «I do not wish to say only cause, but one great final cause . . .» of sexes . . . for otherwise there would be as many species, as individuals, &c . . . few only social . . . hence not social instincts, which as I hope to show is «probably» the foundation of all that is most beautiful in the moral sentiments of the animated beings."[29]

In this intricate cascade of ideas, Darwin traced a path from sexual generation to its consequences: the establishment of stable species, then the appearance of social species, and finally the ultimate purpose of the process, the production of human beings with their moral sentiments. In other words, the end of the process makes intelligible the initial and intermediate stages in the process, indeed, explains their existence. Darwin capped this consideration with a general teleological evaluation that would structure his conception of the final goal of evolutionary nature—man as a moral being: "If man is *one* great object, for which the world was brought into present state . . . & if my theory be true then the formation of sexes rigidly necessary."[30] This particular trajectory needs further explication.

When Darwin opened his first transmutation notebook in spring 1837, he began with his grandfather's reflections on the differences between sexual generation and asexual kinds of reproduction. The grandson supposed that sexually produced offspring would, during gestation, recapitulate the forms of ancestor species. As he initially formulated the principle of recapitulation: "The ordinary kind [i.e., sexual reproduction], which is a longer process, the new individual passing through several stages (typical, ‹of the› or shortened repetition of what the original molecule has done)."[31] Darwin retained the principle of embryological recapitulation right through the several editions of the *Origin*, and thus was in complete accord with his disciple Ernst Haeckel (1834–1919), who made it a central principle of his own science.[32] Recapitulation produced an individual that gathered in itself all the progressive adaptations of its ancestors. But the key to progressive adaptation was the variability that came with sexual reproduction. In spring of 1837, he still

29. Darwin, *Notebook E* (MS pp. 48–49), 409.

30. Ibid.

31. Darwin, *Notebook B* (MS p. 1), 170.

32. I have traced Darwin's development and employment of the principle of embryological recapitulation in Richards, *The Meaning of Evolution: The Morphological Construction and Ideological Reconstruction of Darwin's Theory* (Chicago: University of Chicago Press, 1992), chap. 5. See chapter 6 in the current volume for a discussion of Haeckel's use of the principle of recapitulation.

did not understand exactly how variability might function in adaptation; he yet perceived that variable offspring could adjust to a changing environment in ways that clonally reproducing plants and animals could not. Moreover, in variable offspring, accidental injuries would not accumulate as they would in continuously reproducing asexual organisms. Hence stable species would result from sexual generation. For "without sexual crossing, there would be endless changes . . . & hence there could not be improvement . «& hence not «be» higher animals»."[33] But once stable species were established, social behavior and ultimately moral behavior might ensue.

The idea that Darwin banished final causes from nature, replacing them with mechanistic explanations, obviously cannot be sustained.[34] I have cited only a few instances of several, in the notebooks, of Darwin's use of final causality in the account of natural phenomena. If one adds the many instances in which he employed "purpose"—or its more obscure synonym "object," as in his remark above about man as the "one great object, for which the world was brought into present state"—then both the notebooks and the *Origin* are rife with final-cause language. "Purpose" or "object" occurs some sixty-three times in the *Origin*, while "mechanical," "mechanistic," or any of its forms occurs only five times—and none of them modifying "natural selection." Natural selection hardly operates in Darwin's theory like a Manchester spinning

33. Darwin, *Notebook E* (MS p. 50), 410.

34. Most scholars vigorously assert that Darwin eliminated the metaphysical conceit of teleology from nature. Michael Ghiselin is quite representative. See Ghiselin, "Darwin's Language Might Seem Teleological, but His Thinking Is Another Matter," *Biology and Philosophy* 9 (1994): 489–92. Without bothering to examine Darwin's notebooks, Ghiselin simply asserts: "I have said it before, I will say it again. The notion that Darwin somehow brought teleological thinking back into biology is a myth. In any non-trivial sense of that term, he did the exact opposite. He developed a new way of thinking that allows us to dispense altogether with that metaphysical delusion. I say this not just after having read the whole Darwinian corpus through more than once. Rather, I say it as a professional biologist, who has learned from his own experience, and from that of his colleagues, including Darwin" (489). Ghiselin was reacting to James Lennox, "Darwin *Was* a Teleologist," *Biology and Philosophy* 8 (1993): 409–21. Lennox focuses on the use of final-cause language in Darwin's *Various Contrivances by Which Orchids Are Fertilised by Insects* (1862). Lennox argues that Darwin sought to explain those contrivances by their consequences—that is, by the advantages they exhibited, and thus by natural selection. Lennox's account is quite within the Aristotelian notion of teleology: the final cause, that is, the consequence of a trait or process, illuminates for the biologist the *structure* of the trait or process. In Darwin's analysis, however, the existence of a trait is likely the result of spontaneous variation (or the accumulation of variations), that is, it results from something like an efficient cause in Aristotle's terms. Once in existence a trait, if of advantage, is simply not eliminated. I would side with Lennox over Ghiselin, but I believe there is yet a more fundamental notion of teleology at work in Darwin's theory, which is rather more like Kant's notion of teleology, which does require the assumption of an *archetypus intellectus*, a Divine mind, precisely the sort of entity that both Lennox and Ghiselin presumed Darwin had eliminated from biology.

loom but rather like a refined and morally concerned mind, as I'll try to make clear.[35]

The term "final cause" faded in Darwin's constant reworking of his theory over the two decades prior to the publication of the *Origin*, but the concept remained, supplying support to the whole of his argument. If one does a kind of archeological dig down through the principal documents charting the growth of the theory—from the *Origin*, back through the *Big Species Book* (the manuscript that gave birth to the *Origin*), to the essays of 1844 and 1842, and finally the notebooks—the intellectual layers reveal the structuring work of that teleological conception. So consider the strata underlying the conclusion drawn in the last paragraph of the *Origin*:

1. 1838 (*Notebook E*): "man is one great object, for which the world was brought into present state."[36]
2. 1842 (Essay of 1842): "the highest good, which we can conceive, the creation of the higher animals has directly come."[37]
3. 1844 (Essay of 1844): "the most exalted end which we are capable of conceiving, namely, the creation of the higher animals, has directly proceeded."[38]
4. 1859 (*Origin*): "the most exalted object, which we are capable of conceiving, namely, the production of the higher animals, directly follows."[39]

There is one use of "final cause" that Darwin does repudiate: when a purposive trait is ascribed to the direct action of the Deity instead of to the operations of natural law. In *Notebook M*, Darwin observed: "The unwillingness to consider Creator as governing by laws is probably that as long as we consider each object an act of separate creation, we admire it more, because we can compare it to the standard of our own minds, which ceases to be the case when we

35. Even a shrewd historian such as Gregory Radick easily falls into the locution of mechanism when referring to the way natural selection operates to produce adaptations. See Gregory Radick, "Is the Theory of Natural Selection Independent of Its History," in Hodge and Radick, *Cambridge Companion to Darwin*, 147–72.

36. Darwin, *Notebook E* (MS p. 49), 409.

37. Charles Darwin, "Essay of 1842," ibid., 52.

38. Charles Darwin, "Essay of 1844," in *Foundations of the Origin of Species*, ed. Francis Darwin (Cambridge: Cambridge University Press, 1909), 254.

39. Darwin, *Origin of Species*, 490.

consider the formation of the laws invoking laws, & giving rise at last even to the perception of a final cause."[40]

From the beginning of his theorizing, Darwin argued that events in nature had to be understood as occurring through natural law. But exactly how does that presumption square with his general teleological conception?

## NATURAL SELECTION AS DESIGNED LAW

In the last paragraph of the *Origin of Species*, Darwin specified by way of summary the laws that he discriminated in his book. They included "Growth with Reproduction," "Inheritance," "Variability," "Struggle for Life," and "Natural Selection." Today, we would not likely refer to natural selection as a law but rather as shorthand for sufficient causal forces operative on an organism at a particular time. Darwin, however, in his nineteenth-century way, thought of natural selection as comparable to the law of gravity. In his *Autobiography*, he contrasted his law of natural selection with Paley's intervening deity: "The old argument of design in nature, as given by Paley, which formerly seemed to me so conclusive, fails, now that the law of natural selection has been discovered. We can no longer argue that, for instance, the beautiful hinge of a bivalve shell must have been made by an intelligent being, like the hinge of a door by man."[41] In his notebooks and in the *Origin*, Darwin would contend that the creation of new species occurred by law, the law of natural selection. But what did he mean by law, and how did natural selection operate as a law?

By law, Darwin seems to have meant causal interactions in the natural world that were fixed and of an unchangeable type. These interactions formed a network of radiating forces that governed all organic and inorganic formations. The most general physical causes—for example, slow geological changes—had a determining impact, he suggested, on a more specific range of causes, and these in turn were translated into environmental alterations that caused variations for organic adaptations. He considered this conception of a network of laws shaping organisms to be superior to the "cramped imagination that God created (warring against those very laws he established in all organic nature) the Rhinoceros of Java & Sumatra, that since the time of the Silurian, he has made a long succession of vile Molluscous animals."[42] Darwin's conception of a universe of fixed forces determining all events and even human behavior

40. Charles Darwin, *Notebook M* (MS p. 154), in *Charles Darwin's Notebooks*, 559.

41. Darwin, *Autobiography*, 87. Darwin remembered his Paley imperfectly; the divine also thought the Creator worked through natural laws.

42. Darwin, *Notebook M* (MS pp. 36–37), 343.

seems to have been a significant condition for the rise of the disenchanted modern world. And if one relied on the way both colleagues and enemies characterized his theory, this impression would be correct. This supposition, though, is mitigated both by his assumption concerning the ultimate cause of law itself and by his conception of the operations of the principal law of organic life, natural selection.

As the passage just quoted suggests, Darwin assumed a view of natural law quite common in the early nineteenth century, namely, that law by its very nature required a mind to formulate it and provide the power to enforce it. William Paley, in his *Natural Theology*, expressed this general view: "A law presupposes an agent, for it is only the mode according to which an agent proceeds; it implies a power, for it is the order according to which that power acts. Without this agent, without this power, which are both distinct from itself, the 'law' does nothing; is nothing."[43]

William Whewell (fig. 2.4), whose *History of the Inductive Sciences* Darwin read shortly after his return from the *Beagle* voyage, made a comparable assumption, which for him meant that natural law could be assigned the creative activity in nature; it could act as a surrogate for God. Whewell put it this way in his *Bridgewater Treatise*, in a passage Darwin used as an epigram for the *Origin of Species*: "But with respect to the material world, we can at least go so far as this—we can perceive that events are brought about not by insulated interpositions of Divine power, exerted in each particular case, but by the establishment of general law."[44]

Like Whewell, Darwin believed that the creative power of nature, and thus the explanatory power, lay in natural law. In the manuscript of the *Origin of Species*, Darwin simply defined nature as "the laws ordained by God to govern the Universe."[45] And as he put it to Asa Gray (fig. 2.5), his supporter in America: "I am inclined to look at everything as resulting from designed laws, with the details whether good or bad, left to the working out of what we may call chance."[46] This was no sop to Gray, an adept botanist and even more adroit clergyman. As Darwin confessed in his *Autobiography*, when he wrote the *Origin*, he believed in "a First Cause having an intelligent mind in some degree

43. William Paley, *Natural Theology* (London: Faulder, 1809), 416.

44. William Whewell, *Astronomy and General Physics Considered with Reference to Natural Theology* (Bridgewater Treatise) (Philadelphia: Carey, Lea and Blanchard, 1833), 267.

45. Charles Darwin, *Charles Darwin's Natural Selection, being the Second Part of His Big Species Book Written from 1856 to 1858*, ed. R. C. Stauffer (Cambridge: Cambridge University Press, 1975), 224.

46. Darwin to Asa Gray (22 May 1860), in *The Correspondence of Charles Darwin*, ed. Frederick Burkhardt et al., 19 vols. to date (Cambridge: Cambridge University Press, 1985–), 8:224.

FIGURE 2.4 William Whewell (1794–1866). Lithograph. (© National Portrait Gallery)

analogous to that of man."[47] There seems no good reason to doubt that Darwin was sincere when he contended that God created through secondary causes; it had been his conviction from the first period of his theorizing. His former teacher John Henslow, in a public forum, defended Darwin precisely as imputing to the Creator the ultimate power in the operations of natural law. Henslow

47. Darwin, *Autobiography*, 92–93.

FIGURE 2.5 Asa Gray (1810–1888). Photo.
(Courtesy of South Caroliniana Library)

described this defense in a letter to Joseph Hooker, also a close friend of Darwin's; he said that he had refused "to allow that he [Darwin] was guided by any but truthful motives, and [declared] that he himself believed he was exalting & not debasing our views of a Creator, in attributing to him a power of imposing laws on the Organic World by which to do his work."[48] What Darwin rejected during the period of the composition of his theory was not the notion that

48. John Henslow to Joseph Hooker (10 May 1860), in *Correspondence of Charles Darwin*, 8:200.

God had designed the world for man but that this design should be the handiwork of a tinkering Deity, a God who acted like an English joiner, cobbling the structure of nature ad hoc. Following Whewell, he maintained that the world had to be understood as the product of creative law.[49]

By the time he wrote Gray in spring of 1860, however, Darwin had begun to waiver in his conviction that natural law required an independent designing mind to provide its force. And by the end of the 1860s, he seems to have abandoned altogether the idea that God was a necessary foundation for his theory. What he never abandoned, however, was the ascription to natural selection itself of those properties of discrimination, power, and moral concern previously conferred on it by divine agency. These properties allowed the law of natural selection to lead to the end Darwin foresaw as the goal of the evolutionary process, an outcome that Whewell thought impossible in natural science and rather a conclusion that could be drawn only from theology, namely, the creation of man as a moral creature.

## NATURAL SELECTION AS AN INTELLIGENT AND MORAL FORCE

At the end of October 1838, at the time he considered the "great final cause" of sexual generation—namely, the production of higher animals with their moral traits—Darwin opened his *Notebook N*, in which he began to compose an account of the moral sentiments. He worked out the kernel of his conception, which would later flower in the *Descent of Man*, in a fanciful example. He imagined the case of a dog with incipient moral instincts:

> Dog obeying instincts of running hare is stopped by fleas, also by greater temptation as bitch. . . . Now if dogs mind were so framed that he constantly compared his impressions, & wished he had done so & so for his interest, & found he disobeyed a wish which was part of his system, & constant, for a wish which was only short & might otherwise have been relieved, he would be sorry or have troubled conscience—therefore I say grant reason to any animal with social & sexual instincts «& yet with passions» he must have conscience—this is capital view.—Dogs conscience would not have been same with mans because original instinct different.[50]

49. See John Brooke's masterful discussion "'Laws Impressed on Matter by the Creator'? The *Origin* and the Question of Religion," in *Cambridge Companion to the "Origin of Species,"* ed. Michael Ruse and Robert J. Richards (Cambridge: Cambridge University Press, 2009), 256–74.

50. Charles Darwin, *Notebook N* (MS pp. 1–3), in *Charles Darwin's Notebooks*, 563–64.

Darwin believed that the moral instincts were essentially persistent social instincts that might continue to urge cooperative action even after being interrupted by a more powerful, self-directed impulse. As he suggested to himself at this time: "May not moral sense arise from our enlarged capacity ‹acting› «yet being obscurely guided» or strong instinctive sexual, parental & social instincts give rise 'do unto others as yourself,' 'love thy neighbour as thyself.' Analyse this out."[51] He would indeed continue to analyze out his theory, for at this point in its development he did not see how other-directed, social instincts, which gave no benefit to their carrier, could be produced by selection. This difficulty seems to have led him to retain the device of inherited habit to explain the origin of the social instincts. Thus in late spring 1839, he formulated what he called the "law of utility"—derived from Paley—which supposed that social utility would lead the whole species to adopt certain habits that, through dint of exercise, would become instinctive: "On Law of Utility Nothing but that which has beneficial tendency through many ages would be acquired [i.e., necessary social habits]. . . . It is probable that becomes instinctive which is repeated under many generations."[52] While Darwin never gave up the idea that habits could become inherited, he would solve the problem of the natural selection of social instincts only in the final throes of composing the *Origin*.

Darwin thus looked upon moral impulses as acquired during the course of animal development—not directly implanted in a soul by God. Such moral capacity developed along with more complex brain matter, much in the way the power of gravity became palpable with the increase of mass and as a force intrinsic to it. If mental processes, moral ability, were assigned to matter, would this, however, not be atheism, and thus justify the utter rejection that had already met Lamarck's theory? Darwin didn't think so. As he considered the subject, he bound his kind of materialism into an ennobling teleological framework: "This Materialism does not tend to Atheism. Inutility of so high a mind without further end just same argument. Without indeed we are step towards some final end.—production of higher animals—perhaps, say attribute of such *higher* animals may be looking back. Therefore consciousness, therefore reward in good life."[53]

Darwin here contended that his view of brain-mind did not lead to atheism because the sort of material that produced mind had the final purpose of

51. Darwin, *Notebook M* (MS p. 150), 558.
52. Charles Darwin, *Old and Useless Notes* (MS pp. 50–51), in *Charles Darwin's Notebooks*, 623.
53. Ibid., (MS p. 37), 614.

generating the higher animals, that is, organisms with consciousness, moral standing, and thus the capacity for leading a good life with its (eternal?) reward. As he would put it a few years later, in his essays of 1842 and 1844, this transmutational process led to "the most exalted end which we are capable of conceiving, namely the creation of the higher animals."[54] That, of course, was also the trajectory he specified in the last paragraph of the *Origin of Species.* From the beginning of his career to the publication of the *Origin*, the teleological goal of nature, as his theory construed it, was the production of human beings with their moral sentiments.

At the very end of October 1838, Darwin gave an analytic summary of his developing theory, a neat set of virtually axiomatic principles:

> Three principles, will account for all
> (1) Grandchildren. like grandfathers
> (2) Tendency to small change . . .
> (3) Great fertility in proportion to support of parents.[55]

These factors may be interpreted as follows: traits of organisms are heritable (with occasional reversions); traits vary slightly from generation to generation; and reproduction outstrips food resources (the Malthusian factor). These principles seem very much like those "necessary and sufficient" axioms advanced by contemporary evolutionary theorists: variation, heritability, and differential survival.[56] Such analytic reduction appears to render evolution by natural selection a quite simple concept, as Huxley supposed. These bare principles, however, do not identify a causal force that might scrutinize the traits of organisms to pick out just those that could provide an advantage and thus be preserved. Darwin would soon construct that force as both a moral and an intelligent agent, and the structure of that conception would sink deeply into the language of the *Origin*.

In 1842, Darwin roughly sketched the outlines of his theory, and two years later he enlarged the essay to compose a more complete and systematic version. In the first section of both essays, as in the first chapter of the *Origin*, he discussed artificial selection. He suggested that variations in traits of plants and animals were the result of the effects of the environment, both directly,

54. Darwin, "Essay of 1844," 254. This line, of course, occurs with slight alteration both in the essay of 1842 and in the *Origin of Species*.

55. Darwin, *Notebook E* (MS p. 58), 412–13.

56. See, for example, Richard Lewontin, "Adaptation," *Scientific American* 239 (1978): 212–28.

on features of the malleable body of young progeny, and also indirectly, by the environment's effect on the sexual organs of the parents.[57] Typically a breeder would examine variations in plant or animal offspring; if any captured his fancy, he would breed only from those suitable varieties and prevent backcrosses to the general stock. Backcrosses, of course, would overwhelm or swamp out any advantages that the selected organisms might possess.

In the next section of the essays, Darwin inquired whether variation and selection could be found in nature. Variations in the wild, he thought, would occur much as they did in domestic stocks. But the crucial, two-pronged issue was this: "Is there any means of selecting those offspring which vary in the same manner, crossing them and keeping their offspring separate and thus producing selected races [?]"[58] The first of these problems may be called the problem of *selection*; the second, keeping the selected organisms separate, the problem of *swamping out.* In beginning to engage with these difficulties (and more to come), Darwin proposed to himself a certain model against which he would construct his device of natural selection. This model would control his language and the concepts deployed in the *Origin*. In the 1844 essay, he described the model this way:

> Let us now suppose a Being with penetration sufficient to perceive the differences in the outer and innermost organization quite imperceptible to man, and with forethought extending over future centuries to watch with unerring care and select for any object the offspring of an organism produced under the foregoing circumstances; I can see no conceivable reason why he could not form a new race (or several were he to separate the stock of the original organism and work on several islands) adapted to new ends. As we assume his discrimination, and his forethought, and his steadiness of object, to be incomparably greater than those qualities in man, so we may suppose the beauty and complications of the adaptations of the new races and their differences from the original stock to be greater than in the domestic races produced by man's agency.[59]

The model Darwin had chosen to explain to himself the process of selection in nature was that of a powerfully intelligent being, one that had foresight and

57. Darwin, "Essay of 1842," 1–2.
58. Ibid., 5.
59. Darwin, "Essay of 1844," 85.

selected animals to produce beautiful and intricate structures. This prescient being made choices that were "infinitely wise compared to those of man."[60] As a wise breeder, this being would prevent backcrosses of his flocks. Nature, in the guise of this being, was thus conceived not as a machine but as a supremely intelligent force. In the succeeding sections of the essays of 1842 and 1844, Darwin began specifying the analogs for the model, that is, those features of nature that operated in a fashion comparable to the imaginary being. He stipulated, for instance, that variations in nature would be slight and intermittent due to the actions of a slowly changing environment. But, looking to his model, he supposed that nature would compensate for very slowly appearing minute variations by acting in a way "far more rigid and scrutinizing" than man could execute.[61] He then brought to bear the Malthusian idea of geometrical increase of offspring, and the consequent struggle for existence that would cull all but those having the most beneficial traits.

Many difficulties in the theory of natural selection were yet unsolved in the essays. Darwin had not really dealt with the problem of swamping. Nor had he succeeded in working out how nature might select social, or altruistic, instincts, the ultimate goal of evolution. As he considered the operations of natural selection, it seemed improbable that it could produce organs of great perfection, such as the vertebrate eye. His strategy for solving this last problem, however, seemed ready to hand—namely, to find a graduation of structures in various different species that would illustrate how organs such as the eye might have evolved over long periods of time. Moreover, if natural selection had virtually preternatural discernment, it could operate on exquisitely small variations to produce something as intricate as an eye.

## DARWIN'S *BIG SPECIES BOOK*: COMMUNITY SELECTION AND THE MORALITY OF NATURE

In September 1854, Darwin wrote in his pocket diary, "Began sorting notes for Species theory." His friends had urged him not to delay in publishing his theory, lest someone else beat him to the goal. On 14 May 1856, he recorded: "Began by Lyell's advice writing species sketch."[62] By the following fall, the sketch had grown far beyond his initial intention. His expanding composition was to be called *Natural Selection*, though in his notes he referred to it

60. Darwin, "Essay of 1842," 21.

61. Ibid., 9.

62. Charles Darwin, personal journal, MS 34, DAR 158.1–76, Department of Manuscripts, Cambridge University Library.

affectionately as "my Big Species Book." And big it would have been: his efforts would have yielded a very large work, perhaps extending to two or three fat volumes. The writing was interrupted, however, when Lyell's prophesy of someone forestalling him came true. In mid-June 1858, Darwin received the famous letter from Alfred Russel Wallace (1823–1913), then in Malaya, in which that naturalist included an essay that could have been purloined from Darwin's own notebooks. After reassurances from friends that honor did not require him to toss his manuscript into the flames, Darwin compressed that part of the composition already completed and quickly wrote out the remaining chapters of what became the *Origin of Species*.

At the beginning of March 1858, a few months before he received Wallace's letter, Darwin had finished a chapter in his manuscript titled "Mental Powers and Instincts of Animals." In that chapter he solved a problem about which he had been worrying for almost a decade. In his study of the social insects—especially ants and bees—he recognized that the workers formed different castes with peculiar anatomies and instincts. Yet the workers were sterile, and so natural selection could not act on the individuals to preserve in their offspring any useful habits—they had no offspring. How then had these features of the social insects evolved? In a loose note, dated June 1848, in which he sketched out the problem, he remarked, "I must get up this subject—it is the greatest special difficulty I have met with."[63]

Although Darwin had identified the problem many years before, it was only in the actual writing of the *Big Species Book* that he arrived at a solution. He took his cue from William Youatt's *Cattle: Their Breeds, Management, and Disease.*[64] When breeders wished to produce a herd with desirable characteristics, they chose animals from several different family groups and slaughtered them. If one or another had, say, desired marbling, they would breed from the family of the animal with that characteristic. In the *Big Species Book*, Darwin rendered the discovery this way: "This principle of selection, namely not of the individual which cannot breed, but of the family which produced such individual, has I believe been followed by nature in regard to the neuters amongst social insects; the selected characters being attached exclusively not only to one sex, which is a circumstance of the commonest occurrences, but to a peculiar & sterile state of one sex."[65]

63. Charles Darwin, loose note, DAR 76.1–4, Department of Manuscripts, Cambridge University Library.

64. William Youatt, *Cattle: Their Breeds, Management, and Disease* (London: Library of Useful Knowledge, 1834).

65. Darwin, *Big Species Book*, 370.

Darwin thus came to understand that natural selection operated not only on individuals but also on whole families, hives, or tribes. This insight and the expansion of his theory of natural selection would have three important dividends: first, he could overcome a potentially fatal objection to his theory; second, he could exclude a Lamarckian explanation of the wonderful instincts of the social insects—since no acquired habits could be passed to offspring; and finally, his theory of family selection (or community selection as he came to call it) would enable him to solve the like problem in human evolution, namely, the origin of the altruistic instincts. In the *Descent of Man*, Darwin would mobilize the model of the social insects precisely to construct a theory of human moral behavior that contained a core of pure, unselfish altruism—that is, acts that benefited others at cost to self, something that could not occur under individual selection.[66] Hence, the final goal of evolution, as he originally conceived its telic trajectory, could be realized: the production of the higher animals with their moral sentiments. Yet not only did Darwin construe natural selection as producing moral creatures; he conceived of natural selection itself as a moral and intelligent agent.

The model of an intelligent and moral selector, which Darwin cultivated in his earlier essays, makes an appearance in the *Big Species Book*. In the chapter "On Natural Selection," he contrasted man's selection with nature's. The human selector did not allow "each being to struggle for life"; he rather protected animals "from all enemies." Further, man judged animals only on surface characteristics and often picked countervailing traits. He also allowed crosses that reduced the power of selection. And finally, man acted selfishly, choosing only the property that "pleases or is useful to him." Nature acted quite differently: "She cares not for mere external appearances; she may be said to scrutinize with a severe eye, every nerve, vessel & muscle; every habit, instinct, shade of constitution,—the whole machinery of the organization. There will be here no caprice, no favouring: the good will be preserved & the bad rigidly destroyed."[67]

Nature thus acted steadily, justly, and with divine discernment, separating the good from the bad. Nature, in this conception, was God's surrogate, which Darwin signaled by penciling in his manuscript above the quoted passage: "By nature, I mean the laws ordained by God to govern the Universe." As Darwin pared away the overgrowth of the *Big Species Book*, the intelligent and moral

66. See chapter 3 in the current volume.
67. Darwin, *Big Species Book*, 224.

character of natural selection stood out even more boldly in the précis, that is, in the *Origin of Species*.

## NATURAL AND MORAL SELECTION IN THE *ORIGIN OF SPECIES*

In the first edition of the *Origin*, Darwin approached natural selection from two distinct perspectives, conveyed in two chapters whose titles suggest the distinction: "Struggle for Existence" and "Natural Selection" (chaps. 3 and 4). Although their considerations overlap, the first focuses on the details of the operations of selection and the second contains the more highly personified reconceptualization of its activities. In chapter 3, Darwin proposed that small variations in organisms would give some an advantage in the struggle for life. He then defined natural selection: "Owing to this struggle for life, any variation, however slight and from whatever cause proceeding, if it be in any degree profitable to an individual of any species, . . . will tend to the preservation of that individual, and will generally be inherited by its offspring. The offspring, also, will thus have a better chance of surviving. . . . I have called this principle, by which each slight variation, if useful, is preserved by the term Natural Selection."[68]

Darwin explained what he meant by "struggle" a bit later in the chapter, and I discuss that in a moment. Here, I note several revealing features of his definition. First, selection is supposed to operate on all variations, even those produced by the inheritance of acquired characters and not just those that arise accidentally from the environment acting on the sex organs of parents. Second, Darwin believed that virtually all traits, useful or not, would be heritable—what he called the "strong principle of inheritance." Third, although the initial part of the definition indicates it is the individual that is preserved, in the second part it is the slight variation that is preserved—which latter is the meaning of the phrase "natural selection."[69] The passage draws out "the chicken and egg" problem for Darwin: a trait gives an individual an advantage in its struggle, so that the individual is preserved; the individual, in turn, preserves the trait by passing it on to offspring. Finally, the definition looks to the future, when useful traits will be sifted out and the nonuseful extinguished, along with their carriers. In the short run, individuals are preserved; in the long run, it is their

68. Darwin, *Origin of Species*, 61.
69. Ibid., 5, 61, 81.

morphologies that are both perpetuated and slowly changed as the result of continued selection.

"We behold," Darwin observed (using a recurring metaphor), "the face of nature bright with gladness"; we do not, however, see the struggle that occurs beneath her beaming countenance. But what does "struggle" mean, and who are the antagonists in a struggle for existence? Darwin said he meant "struggle" in a "large and metaphorical sense," which, as he spun out his meandering notion, covered three or four distinct meanings.[70] First, an animal preyed upon will struggle with its aggressor. But as well, two canine animals will "struggle with one another to get food and live." Furthermore, struggle can be used to characterize a plant at the edge of the desert: it struggles "for life against the drought." In addition, one can say that plants struggle with other plants of the same and different species for their seeds to occupy fertile ground. These different kinds of struggle, in Darwin's estimation, can be aligned according to a sliding scale of severity. Accordingly, the struggle will move from most to least intense: between individuals of the same variety of a species; between individuals of different varieties of the same species; between individuals of different species of the same genus; between species members of quite different types; and finally, between individuals and climate. These various and divergent meanings of struggle seem to have come from the two different sources for Darwin's concept: Candolle, who proclaimed that all of nature was at war, and Malthus, who emphasized the population consequences of dearth. Today, we would say that struggle—granted its metaphorical sense—properly occurs only between members of the same species in their efforts to leave progeny. Adopting Candolle's emphasis on the warlike aspects of struggle may have led Darwin to distinguish natural selection from sexual selection, the latter of which concerns not a death struggle for existence but a struggle by males for mating opportunities.[71]

In the chapter "Natural Selection" in the *Origin*, Darwin reintroduced the notion of that powerful intelligence from his essays and the *Big Species Book*, even rendering it with a biblical inflexion:

> Man can act only on external and visible characters: nature cares nothing for appearances, except in so far as they may be useful to any being. She can act on every internal organ, on every shade of constitutional difference, on the whole machinery of life. Man selects only for his own good;

70. Ibid., 62, 62–63.
71. Ibid., 88–90.

> Nature only for that of the being which she tends. . . . Can we wonder, then, that nature's productions should be far "truer" in character than man's productions; that they should be infinitely better adapted to the most complex conditions of life, and should plainly bear the stamp of far higher workmanship? It may be said that natural selection is daily and hourly scrutinizing, throughout the world, every variation, even the slightest; rejecting that which is bad, preserving and adding up all that is good; silently and insensibly working whenever and wherever opportunity offers, at the improvement of each organic being.[72]

The biblical coloring of Darwin's text is condign for a nature that is the divine surrogate and that acts only altruistically for the welfare of creatures, unlike man who acts only for himself. That benevolence extended to every organism, since natural selection worked for "the improvement of each organic being." In the penultimate paragraph of the *Origin*, Darwin again affirmed the moral concern that natural selection evinced: "And as natural selection works solely by and for the good of each being, all corporeal and mental endowments will tend to progress towards perfection." These are not slips of the pen, since he made the same assertion several times throughout the book.[73] But, of course, from our perspective, natural selection does not work for the good of each being. It eliminates most beings; it destroys them. I believe Darwin's conception of a benevolent mind operating in nature had such deep roots in his theory that it overcame what appears to be, at least for us, an obvious consequence of the actions of natural selection—death and extirpation of creatures. In those brief moments when the patent logic of the situation did hit him, he found ways to assuage the consequences: "When we reflect on this struggle, we may console ourselves with the full belief, that the war of nature is not incessant, that no fear is felt, that death is generally prompt, and that the vigorous, the healthy, and the happy survive and multiply."[74]

Even here, Darwin suppressed what he had otherwise maintained, that natural selection is "daily and hourly scrutinizing throughout the world every variation." Natural selection did act constantly; the war of nature was incessant.[75] But Darwin's model of moral agency mitigated the force of Malthusian pitilessness and the implications of his own device.

72. Ibid., 83–84.
73. Ibid., 489. In addition to the passages mentioned, see also 83, 149, 194, and 201.
74. Ibid., 78.
75. Ibid., 84.

## CONCLUSION

I have argued that Darwin did not come to his conception of natural selection in a flash that yielded a fully formed theory. What appears as the intuitive clarity of his device is, I believe, deceptive. I have tried to show that his notions about the parameters of natural selection, what it operates on and its mode of operation, gradually took shape in his mind and hardly came to final form even with the publication of the first edition of the *Origin of Species*. In outlining this gradual evolution of a concept—actually a set of concepts—I have emphasized the way Darwin characterized selection as a moral and intelligent agent. Most contemporary scholars have described Darwinian nature as mechanical, quite amoral in its ruthlessness. To be sure, when Wallace and others pointed out what seemed the misleading implications of the device, Darwin protested that, of course, he did not mean to argue that natural selection was actually an intelligent or moral agent. And by the time of his exchange with Wallace on the subject (1866), he had abandoned any assumption of Divine superintendence.[76] But even Darwin recognized, if dimly, that his original formulation of the device and the cognitively laden language of his writing carried certain consequences with which he did not wish to dispense—and, indeed, could not do so without altering his deeper conception of the character and goal of evolution. Darwin's language and metaphorical mode of thought gave his theory a meaning resistant to any mechanistic interpretation and unyielding even to his later, more cautious reflections.

My analysis depends on recognizing the way teleological conceptions molded Darwin's theory. The discriminating reader may find two conceptions of teleology afoot in Darwin's notebooks and essays, as well as in the *Origin*: one that the modern biologist might tolerate, the other that only the nineteenth-century biologist—at least in Britain—might find satisfactory. The first would be compatible with Aristotle's conception of teleology: granted that human beings now exist, what were the necessary antecedent steps that made their evolution possible? In this consideration, the end—human beings with their various features—would illuminate for the biologist just those determining earlier stages that gave rise to such creatures and, in that sense, would

76. Wallace chided him for the term "natural selection" since it suggested "an intelligent chooser was necessary." See Alfred Russel Wallace to Darwin (2 July 1866), in *Correspondence of Charles Darwin*, 14:228. Darwin had already begun to back away from the apparently intentional character of natural selection in the third edition (1861) of the *Origin*, where he corrected the misapprehension that natural selection expressed "an active power of Deity." See Charles Darwin, *The Origin of Species: A Variorum Text*, ed. Morris Peckham (Philadelphia: University of Pennsylvania Press, 1959), 165.

be a condition for understanding the process (Aristotle's original meaning of cause). The second kind of teleology, however, is usually the one that most scholars think Darwin rejected, namely, that processes in nature unfolded according to a plan. Yet the language of "designed laws," which Darwin explicitly invoked, indicates that the second meaning of teleology was also operative in the construction of his theory. Moreover, the moral solicitude with which natural selection acted and its inevitable progressivist consequences—these must lead, at least they did so for Darwin, to the most exalted object we were capable of conceiving, namely, the production of the highest animals, human beings with their moral instincts.

The theory of evolution by natural selection, embodied in the language and text of the *Origin of Species*, gives no succor to those scholars who would make Darwin's theory theologically and morally neutral. Elliott Sober, for instance, argues that Darwin practiced a kind of methodological naturalism of the sort appropriate for a good scientist today. Sober certainly recognizes Darwin's assertions that natural law stemmed from the Divine mind; he does not, however, appreciate the consequences of that view, which render nature morally saturated and directed to a definite goal. Sober attempts to exculpate Darwin's theory of supernatural taint by claiming that the English master's explanatory appeal to God as a first cause was an "argument for the existence of God," which was a *philosophical* use of his scientific theory; the notion of God as primary cause didn't penetrate or shape the science itself.[77] Now this analysis might save Darwin's conception for contemporary delectation, but it certainly misconstrues the theory as presented in the *Origin*. Darwin was not, pace Sober, demonstrating God's existence; he was assuming it and drawing on the traditional conceptions of God's benevolence and design for nature. Sober has imposed a contemporary construction to obscure the language of Darwin's text and its underlying logic.

Let me spell out some of the more specific consequences of my analysis to make clear how markedly Darwin's original notion of evolution by natural selection differs from what is usually attributed to him. Natural selection, in Darwin's view, moved very slowly and gradually, operating at a stately Lyellian pace (perhaps seizing on useful variations that might occur only after thousands of generations).[78] It compensated for meager variability by daily and hourly scrutinizing every individual, for even the slightest and most obscure variation, to

77. Elliott Sober, "Darwin and Naturalism," in Sober, *Did Darwin Write the Origin Backwards? Philosophical Essays on Darwin's Theory* (Amherst, NY: Prometheus Books, 2011), 121–52, quotation at 128.

78. Darwin, *Origin of Species*, 80, 82.

select just those that gave the organism an advantage.[79] A nineteenth-century machine could not be calibrated to operate on such small variations or on features that might escape human notice. If natural selection clanked along like a Manchester spinning loom, one would not have fine damask—only a skillful and intelligent hand could spin that—or the fabric of the eye.

Second, Darwin frequently remarked in the *Origin* that selection operated more efficiently on species with a large number of individuals in an extensive, open area—what today we call *sympatric speciation*.[80] He presumed that, as in the case of the human breeder, a large number of individual animals or plants would produce more favorable variations upon which selection might act. He had in mind the successful artificial breeders, who kept large flocks, as opposed to the less successful, who had only small stocks from which to breed. Yet in the wild, this scenario for selection could only occur if the watchful eye of an intelligent selector somehow gathered the favored varieties together and isolated them so as to prevent backcrosses into the rest of the stock. When Fleeming Jenkin, in his review of the *Origin*, pointed out the problem of swamping of single variations, Darwin suggested in the fifth edition of the *Origin* that groups of individuals would all vary in the same way due to the impact of the local environment.[81] Thus when the implications of his model of intelligent nature were recognized, Darwin had to invoke as analogue a Lamarckian scenario. Today, we assume that small breeding groups isolated by physical barriers would more likely furnish the requisite conditions for natural selection, thus *allopatric speciation*.

Third, a wise selector that has the good of creatures at heart would produce a progressive evolution, one that created ever more improved organization, which Darwin certainly thought to be the case. He believed that more recent creatures had accumulated progressive traits and would triumph over more ancient creatures regardless of the environments in which they might compete.[82] He summed up his view in the last section of the *Origin*: "And as natu-

79. This is one way of reading what seem contradictory statements in the *Origin*: on the one hand, variations occur only occasionally and at great intervals; on the other, that variations are constant and selection is ways adding them up. In chapter 3, I will suggest that these different conceptions about variations in nature and the pace of selection are an index of the long period over which Darwin constructed the argument of the *Origin*—a period during which he altered his view about the source of variations and about the operations of natural selection. Darwin himself inattentively included these different and warring conceptions in his book.

80. Darwin, *Origin of Species*, 41, 70, 102, 105, 125, 177, 179.

81. Darwin, *Origin: Variorum Text*, 179.

82. Darwin, *Origin of Species*, 336–37.

ral selection works solely by and for the good of each being, all corporeal and mental endowments will tend to progress towards perfection."[83]

Fourth, such an intelligent agent would not merely select for each creature's good but also for that of the community. Darwin, in the fifth and sixth editions of the *Origin*, extended his model of family selection to one that operated simply on a community: "In social animals it [natural selection] will adapt the structure of each individual for the benefit of the community; if this in consequence profits by the selected change."[84]

Finally, the intelligent and moral character of natural selection would produce the goal that Darwin had sighted early in his notebooks, namely, the production of the higher animals with their moral sentiments. Darwin thus concluded his volume with the Miltonic and salvific vision that he harbored from his earliest days: "Thus, from the war of nature, from famine and death, the most exalted object which we are capable of conceiving, namely, the production of the higher animals, directly follows."[85] Darwin's vision of the process of natural selection was anything but mechanical and brutal. Nature, while it may have sacrificed a multitude of its creatures, did so for the higher "object," or purpose, of creating those beings having a moral spine—out of death came life more abundant. We humans, Darwin believed, were the goal of evolution by natural selection. There was indeed "grandeur in this view of life."

## APPENDIX 1. THE LOGIC OF DARWIN'S LONG ARGUMENT

Scholars commonly distinguish two separate conceptions that fly under the rubric "Darwin's theory": common descent of species with modification and natural selection as the causal means by which descent occurs. Ernst Mayr, one of the architects of the modern synthesis, thought that the master himself misled his readers by referring to both of these "very different and independent theories" under the simple designation of "my theory."[86] More recently Elliott Sober also contends that Darwin advanced two logically separate theories in the *Origin of Species*. He maintains that Darwin argued for natural selection in the first part of the book, perhaps for pedagogical reasons, and for common descent only in the later parts of the book. He thus wrote the *Origin*

83. Ibid., 489.

84. Darwin, *Origin: Variorum Text*, 172.

85. Darwin, *Origin of Species*, 490.

86. Ernst Mayr, *One Long Argument: Charles Darwin and the Genesis of Modern Evolutionary Theory* (Cambridge: Harvard University Press, 1991), 36.

backwards.[87] This paradoxical claim depends on the further assertion that the epistemic logic of the relationship entails that common descent ought to be established first and evidence for its causal structure be given only thereafter. Mayr, Sober, and other scholars have failed to take Darwin's own observation about his accomplishment seriously, namely, that his "whole volume is one long argument."[88]

I will focus on Sober's analysis since he makes articulately evident what others have simply assumed; he shows what is at stake in maintaining the logical independence of the conceptions of common ancestry and of natural selection. His argument has two parts, one quasi-empirical, about the actual structure of Darwin's book, and one logical, about the implicative relationships governing that structure. He maintains that Darwin "front-loads his discussion of natural selection and lets his full argument for common ancestry emerge only later and in somewhat fragmented form"; he suggests those arguments for the genealogical descent of species come principally in chapter 13, which deals with classification, morphology, embryology, and rudimentary organs.[89] Darwin intended to show, according to Sober, how these various approaches revealed similarities among species, thus providing evidence for common descent. Sober's assertion about the general structure of Darwin's book—that he delayed discussion of common descent to the last part of his treatise—is, on the surface, implausible, as a brief overview of the chapters makes clear.

Chapter 1, "Variation under Domestication," discusses the descent of various races of domestic animals from common ancestral forms. Chapter 2, "Variation in Nature," argues that there is no real difference between varieties and species and that patterns of their relationship, as described by naturalists, provide evidence of species descent from a common ancestor. The second part of chapter 4, "Natural Selection," details Darwin's principle of divergence, which explains the kind of branching characteristic of phylogenetic descent. Chapter 5, "Laws of Variation," specifies evidence of similar variability to argue for common descent. Chapter 6, "Difficulties on the Theory," sketches the wonderfully imaginative possibilities of common descent: whales from bears and birds from flying fish. Chapter 7, "Instinct," argues for the likely descent of the honey bee from something like the Mexican *Melipona*, and they from something like the humble bee. Chapters 9 and 10 on the geological record are, of course, all about common descent from evidence of the fossil record. Chap-

87. Sober, *Did Darwin Write the Origin Backwards?* chap. 1.

88. Darwin, *Origin of Species*, 459.

89. Sober, *Did Darwin Write the Origin Backwards?* 33.

ters 11 and 12 explore biogeographical relationships—particularly how it can occur that species of a common genus can be found at significant geographical distances from each other, though they "must originally have proceeded from the same source, as they have descended from the same progenitor."[90] Arguments about descent with modification thus occupy three of the first four chapters of Darwin's book and can be found throughout virtually every chapter thereafter. Moreover, natural selection runs in tandem with arguments for common ancestry through most of the chapters of the book. Darwin structured his chapters this way for good, logical reasons. An examination of the first three chapters of the *Origin* will bring into relief the logical connections he established between common descent and his device of natural selection.

Chapter 1 of the *Origin* deals with what we call "artificial selection." Darwin believed that once his readers understood the process by which breeders actually produced domestic stocks, he would have a persuasive analogy for selection in nature. Lamarck had used the breeder's selection as a model for his theory of the descent of species. Lyell countered that artificial selection could not act as a model for species descent, since savages originally had chosen as domestic animals those that were morphologically malleable—thus we should not expect to find such animals in nature.[91] Darwin attempted to nullify Lyell's argument in two ways. He first pointed out that savages could not have known antecedently which animals would prove to be more plastic. But this was not a powerful argument, and Darwin knew it. What was powerful was his demonstration that the weird and wonderful array of fancy pigeons—fantails, pouters, nuns, tumblers, and the many others—had a common descent from the ordinary rock pigeon, *Columba livia*. Most breeders had assumed that the various breeds of pigeon had been found originally in nature. If that were so, then artificial selection, as Darwin newly conceived it, could not be shown to be an effective device for transforming organisms, and thus it could not serve as a model for natural selection. So Darwin had to demonstrate common descent from the rock pigeon in order to show the power of artificial selection and to set the grounds for natural selection.

Darwin himself had become a pigeon fancier, raising pigeons by the score in his backyard.[92] He had several arguments for the common descent of fancy pigeons, but the most powerful were his experiments in cross-breeding of varieties to reveal in the offspring strong coloring traits of the rock pigeon. These

90. Darwin, *Origin of Species*, 351.

91. Lyell, *Principles of Geology*, 2:26.

92. I discuss Darwin's pigeon-breeding activities in more detail in chapter 3.

experiments made the case for the effectiveness of the breeder's selection and ultimately of nature's selection. Darwin, however, had another task in this first chapter: he also had to show what artificial selection amounted to so that he might argue persuasively for common descent from the rock pigeon. Exactly what the breeder was doing in producing domestic stocks was little understood when Darwin wrote; his chapter helped clarify the process by which breeders selected variations that happened to occur and cross-bred those animals bearing the favored traits.

From variation in domestication, Darwin turned in chapter 2 to variation in nature. He deployed the term "variation" in different, though related senses: it referred, first, to the way offspring would vary from their parents and from each other, and, second, to the way groups of individuals would vary within a species—that is, would form varieties. And Darwin extended his analysis to the way species would constitute varying forms of a genus, and genera of a family, right up the taxonomic categories. He wanted to demonstrate that these various conceptions of variety carried an implication important for his theory: by the term "variety . . . community of descent is almost universally implied."[93] He devoted the first half of the chapter to showing that no qualitative distinction, only degrees of similarity, distinguished individual differences from varieties, and varieties from species—certainly some experts, he pointed out, would describe as a species what others would describe as a variety. He concluded that "these differences blend into each other in an insensible series; and a series impresses the mind with the idea of an actual passage"—that is, a passage of common descent.[94]

The second part of chapter 2 provides statistical evidence for the kind of descent relations referred to in the first part. Darwin scrutinized some twelve large flora books to discover patterns of descent.[95] He determined that of the possible patterns, his theory predicted the one that was the most prominent, namely, that in large genera (i.e., genera with a large number of species) the species were also large (i.e., had a large number of varieties), supporting the thesis that current species derived from past varieties. Had his statistical analysis shown, say, that large genera had small species (i.e., each with a small number of varieties), his hypothesis would not have been confirmed. Darwin drew a complementary kind of evidence for descent, when he argued on the basis of analogy: "species of large genera present a strong analogy with varieties.

93. Darwin, *Origin of Species*, 44.
94. Ibid., 51.
95. See chapter 3 of this volume for the elaboration of Darwin's statistical arguments.

And we can clearly understand these analogies if species have once existed as varieties, and have thus originated: whereas, these analogies are utterly inexplicable if each species has been independently created."[96] These were ingenious arguments for common descent from patterns of species relationships. But regardless of the validity of his conclusions, it is quite obvious that the second chapter of the *Origin* was devoted to arguments for common ancestry. Sober's belief that Darwin reserved arguments for genealogical descent only to the later chapters of his book simply cannot stand.

Chapter 3 of the *Origin*, "Struggle for Existence," shows what must be the consequence of the kind of variation demonstrated in chapter 2: because of great fecundity of organisms, there must be a struggle for existence among the different creatures within a variety, of different varieties within a species, of different species within a genus, and so on. Thus, without variation in nature, comparable to what the domestic breeder found within his stocks, natural selection could not operate. So the theory of natural selection required, logically required, a demonstration of the kinds of variety and of their relationships that Darwin evinced in chapter 2.

Sober argues that Darwin began the exposition of his theory concentrating on natural selection, giving it causal priority and saving the arguments for descent from a common ancestor until the latter part of his book. If he were writing the *Origin* in an epistemically logical fashion, he would have given, Sober maintains, common descent evidentiary priority—that is, he would have put it at the beginning of his book. I have tried to demonstrate in this appendix that common descent is argued for in the first several chapters of the *Orign*, where it is logically intertwined with the conception of natural selection. Let me close with a crucial logical point.

Sober and other scholars depict Darwin as deploying the kind of abstract considerations that might regard common descent and natural selection as logically distinct: "Darwin's case for common ancestry," Sober declares, "does not depend at all on natural selection's causing evolution."[97] Perhaps this makes sense for a contemporary philosopher of biology. (But even this, I doubt: for if there is a rational or epistemically proper order, the theories of descent and natural selection cannot be logically separate.) In dealing with the historical Darwin, however, we must look at the logic he actually deployed—and therein lies a difference.

96. Darwin, *Origin of Species*, 59.
97. Sober, *Did Darwin Write the Origin Backwards?* 44.

Darwin assumed that his readers would be quite aware of the Linnaean systematic arrangements, as well as those of other systematists; all such schemes were based on affinities or resemblances among the groupings. Darwin did indeed discuss such similarities in chapter 13 of the *Origin* (as well as in many other chapters). But these similarities could not be taken in logical isolation as evidence for common ancestry—and indeed, were not so taken by professional naturalists before Darwin. In every category that Darwin mentioned in chapter 13, the similarities had been noted by many naturalists, and that recognition did not lead them to posit descent from common ancestors, rather to reveal a common plan of the Creator. Similarities could be taken as evidence for common descent only after the introduction of an effective causal device that might explain transitions from one species to another. After Darwin made the case for his causal device, then resemblance, which all zoologists had recognized, could be turned into evidence for common descent—when a natural principle could render resemblance into a process of nature instead of a plan devised by the Creator. Lamarck tried it, but his device seemed ineffective and the range of the evidence insufficient. Darwin, like Lamarck, started with artificial selection, but showed, as his predecessor had not, what was really involved in domestic breeding—how the breeder selected out certain variations, mated their carriers, and continued that process until a different morphological type was produced. He then argued in reverse, as it were: from the cross-breeding of established varieties of fancy pigeons to their common ancestor in the rock pigeon. In the first chapter of the *Origin*, Darwin thus provided mutually implicative arguments for common ancestry and for an effective model of what occurs in nature. This general pattern of argument followed in the rest of the chapters. And that is why both descent and natural selection ran hand in hand throughout the *Origin*—each logically dependent on the other. Darwin had constructed one long, epistemically structured argument.

## APPENDIX 2. THE HISTORICAL ONTOLOGY AND LOCATION OF SCIENTIFIC THEORIES

The question of the location of Darwin's theory may seem anomalous. Yet, we take for granted that his theory exists and that, therefore, it has an ontology, and thus presumably some kind of location. The question of where it exists is not, then, outré, especially if we allow some latitude as to what counts as a possible place. Karl Popper had an answer to the question; he claimed scientific theories existed in a third, quasi-Platonic world, which he distinguished from two other worlds. He conceived the geography in this way: "first, the world of physical objects or of physical states; secondly, the world of states of consciousness, or of mental states, or perhaps of behavioural dispositions to act; and thirdly, the world of *objective contents of thought*, especially of scientific and poetic thoughts and of works of art."[98]

Scientific theories exist, according to Popper, in the same way and in the same abstract manner as Euclid's Pythagorean Theorem. It is what we think about when we think about the Pythagorean Theorem. When Euclid demonstrated that the sum of the areas of squares drawn on the legs of a right triangle was equal to the area of the square drawn on the hypotenuse, he was not proving this of a triangle sketched in the sand or one imagined by himself—strictly speaking, these would fail to be right triangles; he demonstrated the properties of The Right Triangle, the objective triangle of which we might discover yet further attributes. As I suggested in chapter 1, this view of theories, without further qualification, could easily lead to the presumption that a theory like Darwin's might have instantiations in 1859 and today while remaining essentially the same abstract structure, with the further consequence that we might easily read our current notions about evolution back into Darwin's original theory. I believe we are faced with this outcome often in the large literature describing Darwin's accomplishment, bereft as it often is of what it means to be an early nineteenth-century thinker.

We cannot exorcise Plato's ghost completely, however. We would not wish to identify Darwin's theory simply with ink smudges on the paper sitting in John Murray's warehouse or on the paper that holds the reproduction of those smudges in the Harvard University Press paperback. Darwin's theory has a logic and set of implications that transcend ink on paper. Nor does it quite do to presume that the theory existed as a collection of ideas in Darwin's head. Darwin no longer exists, but his theory surely does exist for us; moreover, the theory, at least as most historians would regard it, has depths that might have escaped even Darwin's explicit awareness at any particular time. For example, in his *Autobiography*, Darwin claimed that in the 1830s and early 1840s, he never explicitly formulated a theory but simply collected facts in a wholesale manner. But we see from his notebooks that the phrase "according to my theory" lies scattered through their pages. When he wrote the *Autobiography*, he knew that a good scientist worked in a Baconian manner, collecting facts

98. Karl Popper, "Epistemology without a Knowing Subject," in Popper, *Objective Knowledge: An Evolutionary Approach* (Oxford: Oxford University Press, 1972), 106.

before using them to construct a theory.[99] That presumption falsified his own memory of his achievement.

My analysis in this chapter supposes that the book *Origin of Species* expresses Darwin's theory, points to it, and that the theory as so expressed is the culmination of a development that began at least as far back as the *Beagle* voyage; this developing theory, I believe, has depths not entirely transparent even to Darwin himself. This relative opacity led him later to reject Wallace's suggestion that he drop the phrase "natural selection" and replace it with Spencer's "survival of the fittest."[100] Darwin agreed with Wallace that "survival of the fittest" seemed adequately to perform the function of "natural selection," yet, he demurred. He thought his original expression captured something necessary to his theory, something beyond the substitute suggested by Wallace, though he could not exactly say what that something was. I have urged that the missing features, not quite obvious even to Darwin himself in the mid-1860s, were the intentional and teleological structures that originally came to invest the theory during its early development. Popper was right in this respect: scientific theories have features not simply identified with words on a page. Yet, in Darwin's case, the theory and its logic were generated by the words he jotted in his notebooks and essays, and, of course, by the ideas in his head. That logical structure so generated became a permanent part of Darwin's theory, at least as it existed in 1859, but the theory did continue to evolve through the mid-1860s.

Darwin's theory has an existence comparable to that of a species. We don't identify a species with this or that individual organism or even with the entire group of species members existing at any one time. We don't make this nominalist identification since we typically include as members of the species individuals that no long exist and those that will shortly come to exist. But, of course, there is a further reason for not identifying species with its members. Certain individuals reproductively related to others may not exhibit all of the traits identified with the species—that is, we may attribute features to the species not exactly realized in some of its members; for example, bipedalism may be characteristic of the human species, though there will be members who are, for a number of causes, without lower limbs. Moreover, and this is the primary reason for not identifying a species with its members, species evolve but individuals do not. Evolution is a trait of species but not of individuals or collections of individuals. In this latter respect, we might compare theories

99. Darwin, *Autobiography*, 119.
100. Alfred Russel Wallace to Darwin (2 July 1866).

with species: theories also evolve, though neither the words on a page nor even the individual ideas in the mind of the theorist evolve. I believe we want to say something such as the following: theories have an abstract existence, though generated by individual acts of a theorist and intimately tied to those particular acts.

If theories have this abstract but temporally anchored existence, they transcend the individual instantiations that gave them rise, and this accounts for their public and objective character. They are not simply individual creatures of the theorist's brain. But if they have a public existence, how exactly are they apprehended by the public, that is, by the consumers of theory, including historians? I believe this understanding occurs through a grasp of the words, in their contemporaneous meanings, instantiating the theory. So, when the historian tries to come to terms—literally, come to terms—with, say, Darwin's theory, he or she will construe the meaning of the words Darwin used in the way a mid-nineteenth-century individual of considerable education would. But even beyond that, the exacting historian will take into account the local environment of Darwin's theorizing to determine any inflections of meaning that his particular usage would suggest.

A theory has a transcendent existence, though one continuously generated by the acts of the scientist. Yet because a theory escapes the private realm through the public meaning of words, it has an abstract and objective character, and that character may have features not completely transparent to even its originator. A well-developed scientific theory is like a well-wrought urn: it has a public existence and manifests aesthetic and logical features perhaps unanticipated by the craftsman. Hence it is possible for the historian to say of a theorist like Darwin that he did not fully appreciate his own theory, especially when reflecting back on it at a subsequent time.

A theory with the kind of existence I am suggesting never strays far from the acts that brought it into existence, so it is always located temporally in conjunction with those acts. Yet it continues to evolve, at the hands of both its creator and others who take it up. As it undergoes evolution, much like a species, it is impossible to be precise about exactly when it comes into existence and when it passes into another theory—when Darwinian theory becomes, for example, "neo-Darwinian theory," to use George Romanes's locution, or "ultra-Darwinian theory," to use another of Romanes's formulations.[101] Just so, it is

101. Romanes coined both "neo-Darwinian" and "ultra-Darwinian." See George Romanes, *Darwin and after Darwin*, 4th ed., 3 vols. (Chicago: Open Court, 1916), 2:7, 232. By both designations he principally meant Friedrich Weismann's theory, which precluded Lamarckian devices.

something of an arbitrary decision to mark the temporal joint where therapsids, the mammal-like reptiles, became mammals.

While my analysis here descends into the hazards of metaphysics, most historians remain wary of these vertiginous deeps. Careful historians nonetheless make implicit assumptions that bring them close to the very edge of such considerations.

CHAPTER THREE

# Darwin's Principle of Divergence

## *Why Fodor Was Almost Right*

In a series of articles and in a recent book, *What Darwin Got Wrong*, Jerry Fodor has objected to Darwin's principle of natural selection on the grounds that it assumes nature has intentions.[1] Despite the near universal rejection of Fodor's argument by biologists and philosophers of biology (myself included),[2] I now believe he was almost right. I will show this through a historical examination of a principle that Darwin thought as important as natural selection, his principle of divergence. The principle was designed to explain a phenomenon obvious to any observer of nature, namely, that animals and plants form a hierarchy of clusters. Theodosius Dobzhansky made this the motivating observation of his great synthesizing work, *Genetics and the Origin of Species* (1937): "the living world is not a single array of individuals in which any two variants are connected by a series of intergrades, but an array of more or less distinctly separate arrays, intermediates between which are absent or at least rare. . . . Small clusters are grouped together into larger secondary ones, these into still

1. See, for instance, Jerry Fodor, "Why Pigs Don't Fly," *London Review of Books* 29 (October 2007): 19–22; Jerry Fodor, "Against Darwinism," *Mind and Language* 23 (2008): 1–24; and Jerry Fodor and Massimo Piatelli-Palmarini, *What Darwin Got Wrong* (New York: Farrar, Straus and Giroux, 2010).

2. See, for example, Ned Block and Philip Kitcher, "Misunderstanding Darwin: Natural Selection's Secular Critics Get It Wrong," *Boston Review*, March–April 2010, 29–32; Elliott Sober, "Natural Selection, Causality, and Laws: What Fodor and Piatelli-Palmarini Got Wrong," *Philosophy of Science* 77 (2010): 594–607; Daniel Dennett, "Fun and Games in Fantasyland," *Mind and Language* 23 (2008): 25–31; Peter Godfrey-Smith, "Explanation in Evolutionary Biology," *Mind and Language* 23 (2008): 32–41; and Robert J. Richards, "Darwin Tried and True," *American Scientist* 96 (May–June 2010): 238–42.

larger ones, and so on in a hierarchical order."[3] Nested groupings allow the naturalist to apply the Linnaean taxonomic categories of variety, species, genus, family, and so on. The explanation of divergent clusters remains, however, an area of biology still in dispute. Darwin thought the solution to the problem central to his theory, and he devoted considerable attention to it. His account of divergence presents some quite curious perplexities and illuminates hidden features of his other chief principle, natural selection. Those features have led me to reevaluate Fodor's argument against Darwinian theory.

## DARWIN'S DISCOVERY OF THE PRINCIPLE OF DIVERGENCE

Darwin recalled in his *Autobiography* that a significant problem had escaped his notice during the early 1840s, when he first summarized his theory of species transmutation. His essays of 1842 and 1844 simply failed, he said, to explain the origin of the morphological gaps separating species and the even wider ones among genera and the higher taxa.[4] One can understand why Darwin would have thought the difficulty significant. After all, a theory of the gradual descent of species, with new species slowly emerging from older ones, would seem to forecast smooth transitions among both species and the higher taxonomic groupings, with no missing links. Yet systematic relations among species hardly displayed the expected insensible transitions, even when fossils were brought into the picture. Darwin marked it as the "gravest objection which can be urged against my theory," since it had the power to undermine the basic conception of a gradual evolution of species.[5] Even today religious opponents raise this particular objection with avidity. In the *Autobiography*, Darwin stated the problem and then portrayed his solution as a dramatic, eureka moment:

> At the time [in the mid-1840s], I overlooked one problem of great importance. . . . This problem is the tendency in organic beings descended from the same stock to diverge in character as they become modified.

3. Theodosius Dobzhansky, *Genetics and the Origin of Species*, with an introduction by Stephen Jay Gould (New York: Columbia University Press, [1937] 1982), 4.

4. Darwin's essays of 1842 and 1844 were never published in his lifetime. His son Francis published them on the hundredth anniversary of his father's birth. See Charles Darwin, *Foundations of the Origin of Species: Two Essays Written in 1842 and 1844*, ed. Francis Darwin (Cambridge: Cambridge University Press, 1909).

5. Charles Darwin, *On the Origin of Species* (London: John Murray, 1859), 280.

> That they have diverged greatly is obvious from the manner in which species of all kinds can be classed under genera, genera under families, families under suborders, and so forth; and I can remember the very spot in the road, whilst in my carriage, when to my joy the solution occurred to me; and this was long after I had come to Down. The solution, as I believe, is that modified offspring of all dominant and increasing forms tend to become adapted to many and highly diversified places in the economy of nature.[6]

From his recollection, it appears that the problem and its solution came to him more or less in the same period. The evidence, which I will shortly recount, is otherwise. In any case, the principle of divergence was clearly quite important in Darwin's estimation. He wrote his friend Joseph Hooker in June 1858 that "the 'Principle of Divergence,' . . . along with 'Natural Selection,' is the keystone of my book; and I have very great confidence it is sound."[7]

The earliest explicit mention of the principle came in the large manuscript Darwin had begun in 1856, his "Big Species Book." The writing of that manuscript was interrupted in June 1858 when he received Wallace's letter, which included an essay sketching virtually the very theory of transmutation of species he had been long laboring over. After some encouragement from his friends—he had to be persuaded that he had not lost his originality and that honor did not require him to abandon his manuscript—Darwin abridged the chapters of the *Big Species Book* that he had finished and added others to complete what he called his "abstract." This abstract was published in November 1859 as the *Origin of Species*. Earlier, in March 1857, he had piled up pages of a first draft of chapter 6 of the *Big Species Book*, which touched on divergence; during the next few months, into spring of 1858, he added to the chapter some forty manuscript pages expanding his discussion. That chapter is comparable to chapter 4 of the *Origin*, the second half of which is devoted to the principle of divergence. These dates suggest that the problem of divergence and its solution arose for him in the mid-1850s when he was working on his manuscript. At least by his own testimony, the problem had not occurred to him until after he had written the essay of 1844.

The emphasis that Darwin placed on the late recognition of the problem of divergence and the discovery of its solution is startling. After all, doesn't

6. Charles Darwin, *The Autobiography of Charles Darwin, 1809–1882*, ed. Nora Barlow (New York: Norton, 1969), 120–21.

7. Darwin to Joseph Hooker (8 June 1858), in *The Correspondence of Charles Darwin*, ed. Frederick Burkhardt et al., 19 vols. to date (Cambridge: Cambridge University Press, 1985–), 7:102.

natural selection, in adapting organisms to an environment, competitively separate them to form distinct varieties, and don't these varieties, with further selection, become ever more discrete and therefore morphologically separate species? In other words, natural selection selects differences, and over time these differences naturally become greater in a changing environment, with the result that groups of organisms diverge from one another. Didn't Darwin appreciate this process early in his theorizing? Is a special principle required then to explain divergence?

## WHEN DID DARWIN RECOGNIZE THE PROBLEM OF DIVERGENCE?

Even before he formulated the rudiments of his device of natural selection in late September 1838, Darwin recognized that his emerging theory of branching could explain the applicability of the taxonomic categories. This recognition is depicted in that very early and now famous tree diagram from Darwin's *Notebook B* (see fig. 3.1), which he began during late spring or early summer of 1837.[8]

Beneath the diagram he wrote: "Thus between A & B immense gap of relation. C & B the finest gradation. B & D rather greater distinction. Thus genera would be formed.—bearing relation to ancient types." In the figure, Darwin depicted a remote common ancestor at 1 as ultimately yielding descendant species, which were represented at the ends of branches with terminal crossbars (those without bars indicated extinction). These species were grouped into four genera at nodes standing for the most recent common ancestor: three species at A, four at B, and three at C and D. The nodes at these groupings also denoted the morphological type of the ancestor that gave rise to the species at the branch endings. The splitting branches would produce, as Darwin remarked in his notebook, the morphological gaps among these groups, greater between the genus groupings at A and B, smaller between those at C and B. Although Darwin did not explicitly do so in the notebook, the diagram could also have illustrated other Linnaean categories. The more interior

8. Charles Darwin, *Notebook B* (MS p. 36), in *Charles Darwin's Notebooks, 1836–1844*, ed. Paul Barrett et al. (Ithaca: Cornell University Press, 1987), 180. I believe that Darwin's diagram was inspired by a similar one rendered by the Scots embryologist Martin Barry, who represented in the branching of the several classes of animals Karl Ernst von Baer's theory of development. Barry labeled his diagram "The Tree of Animal Development." I discuss this source in Richards, *The Meaning of Evolution: The Morphological Construction and Ideological Reconstruction of Darwin's Theory* (Chicago: University of Chicago Press, 1992), 108–11.

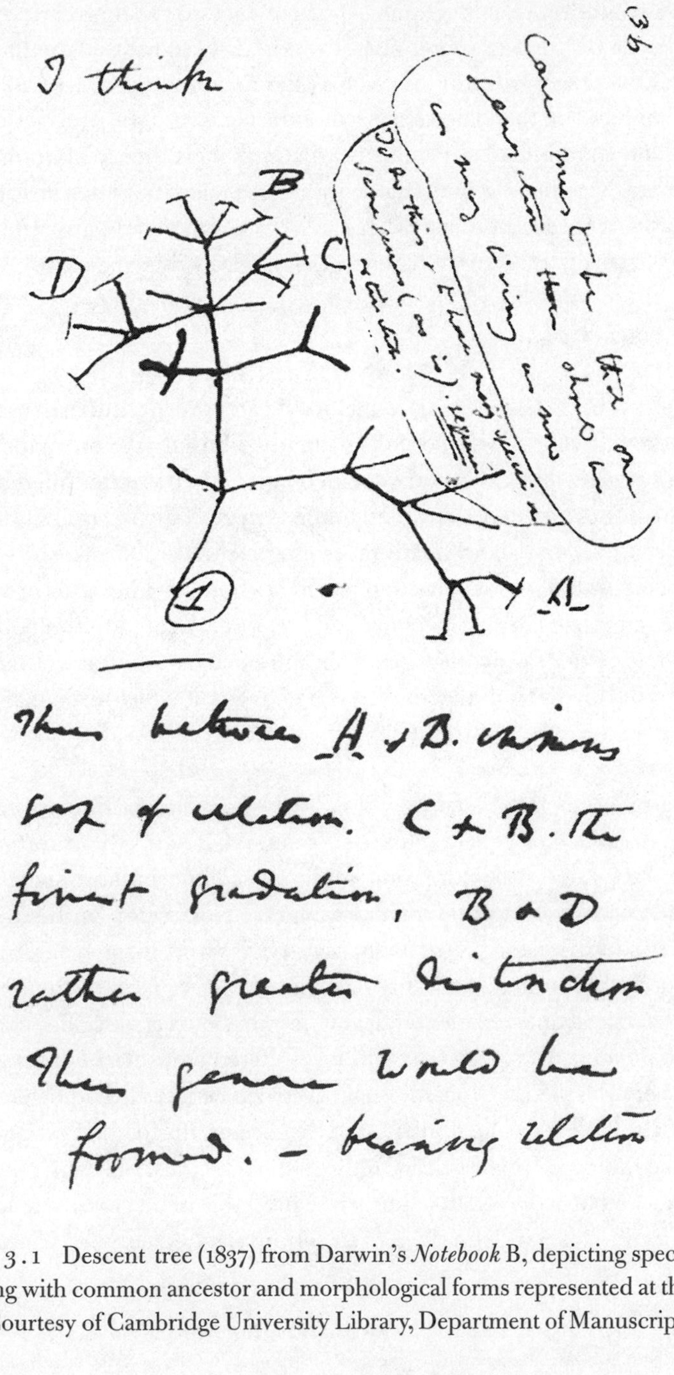


FIGURE 3.1 Descent tree (1837) from Darwin's *Notebook* B, depicting species branching with common ancestor and morphological forms represented at the nodes. (Courtesy of Cambridge University Library, Department of Manuscripts)

nodes would represent still more remote ancestor species. For instance, the next node up from the grouping at A could stand for the ancestor that produced the *genus* group A—as well as the morphological type of the *family*; the first node on the main stem, that of the *class*; and the number 1, that of the *order*. So Darwin had recognized early on that his theory of branching could illustrate the widening gaps among the taxonomic groupings. Perhaps, though, he had not focused on just what caused the branched gaps. In the essay of 1844, however, he seems to have treated precisely this question.

In that essay, Darwin appears to have given an early version of the principle of divergence. He wrote:

> Let us suppose for example that a species spreads and arrives at six or more different regions, or being already diffused over one wide area, let this area be divided into six distinct regions, exposed to different conditions, and with stations slightly different, not fully occupied with other species, so that six different races or species were formed by selection, each best fitted to its new habits and station. . . . The races or new species supposed to be formed would be closely related to each other; and would either form a new genus or sub-genus. . . . In the course of ages and during the contingent physical changes, it is probable that some of the six new species would be destroyed.[9]

Darwin then described how this process would continue; he concluded: "The existence of genera, families, orders, & c., and their mutual relations naturally ensues from extinction going on at all periods amongst the diverging descendants of a common stock."[10] His explanation of the divergence of species in these passages—namely, that species were formed and became morphologically distinct by occupying different places in the economy of nature and that extinctions would delineate the gaps between species—appears to be approximately the same explanation he offered in his *Autobiography* as a new discovery post-1844. What, then, did Darwin believe he had neglected before the 1850s? What did he think he had discovered during his carriage ride?

The foregoing puzzles lead to three specific questions I investigate in this chapter. What is the relationship of the principle of divergence to that of natural selection? Is it independent of selection; is it derivative of selection; or is it

9. Charles Darwin, "Essay of 1844," in *Foundations of the Origin of Species*, 208–9.

10. Ibid., 213. Darwin suggests much the same idea, though in a vague way, in "Essay of 1842," in *Foundations of the Origin of Species*, 36–37.

a type of selection, perhaps comparable to sexual selection? Second: What is the advantage of divergence that the principle implies—that is, why is increased divergence beneficial in the struggle for life? And finally: What led Darwin to believe he had discovered the principle only in the 1850s? The resolution of these questions has implications for Darwin's other principle, natural selection, and for the validity of Fodor's argument dismissing natural selection as a coherent principle of biology.

## DARWIN'S BOTANICAL STATISTICS

The very day, 9 September 1854, after he closed the final volume of his barnacle systematics—four volumes on the known species of barnacles, extant and extinct—Darwin, as he noted in his pocket diary, "began sorting notes for Species Theory."[11] From that time until the fall of 1859, when the *Origin of Species* appeared, he worked steadily on that theory. It was during this concentrated effort that many new ideas emerged, including a fresh set of notions about species divergence.

Darwin began the actual work of composing the *Big Species Book* in May 1856. He discussed the principle of divergence in chapter 6, titled "Natural Selection," which he began writing in early March 1857. Many of the ideas in the chapter, however, took form earlier in the composition, when he was working on variation in nature—in chapter 4, which he began in late December 1856. During this period, Darwin had been inspired to attempt a mathematical demonstration of certain hypotheses about likely patterns of relationship among genera, species, and varieties.[12] He had been aware that botanists had devised ratio calculations to determine, for example, the number of species per family that were indigenous to one region as against the number that were spread over several regions.[13] He also did some preliminary calculations in late 1854

11. The four volumes are Charles Darwin, *Living Cirripedia, a Monograph on the Sub-class Cirripedia, with Figures of All the Species*, vol. 1: *The Lepadidae; or, Pedunculated Cirripedes*, and vol. 2: *The Balanidae, (or Sessile Cirripedes); the Verrucidæ* (London: Ray Society, 1852 and1854); *A Monograph on the Fossil Lepadidae, or, Pedunculated Cirripedes of Great Britain* (London: Printed for the Palaeontographical Society, 1851); and *A Monograph on the Fossil Balanidae and Verrucidae of Great Britain* (London: Printed for the Palaeontographical Society, 1854). Quotation at Charles Darwin, *Personal Journal*, in *Correspondence of Charles Darwin*, 5:537 (Appendix I).

12. Darwin finished a first draft of chapter 4 in January 1857. He added his statistical work in a second draft, completed in April 1858. See Darwin, *Personal Journal*, 6:523 (Appendix II), 7:503 (Appendix II).

13. Alphonse de Candolle performed this kind of calculation over many plant families—that is, for a given family, the ratio of the number of species indigenous to a single region as against the number common to several regions. See especially the second volume of Candolle, *Géographie botanique raisonnée ou exposition des faits principaux et des lois concernant la distribution géographique des plantes de l'époque*

on the ratio of species in so-called aberrant genera (i.e., those hard to place in a particular family) to those in normal genera.[14] With the aid of a schoolmaster whom he hired for the purpose, he went through several large catalogues of the plants found in different countries—for instance, the plants of Great Britain, New Zealand, Russia, and so on—some twelve flora books in all. For each of the catalogues, he counted the number of genera that were large (i.e., had a large number of species) in relation to those that were small.[15] He also tabulated the number of large species (i.e., species with a large number of varieties) compared to the number of small species. He then determined the number of dominate species—that is, species with many individuals spread over several regions of a country—that were found in the large genera as against those in the small. From these tabulations he made a series of statistical judgments. His analyses showed that large genera—that is, those with many species—tended to have large species—that is, species with a large number of varieties.[16] Moreover he found that it was the dominant species that tended both to have a large number of varieties and to be included in the large genera. The numerical evidence thus supported his primary hypothesis, namely, that current species were originally varieties of earlier species.[17] Had he found that small genera tended to have large species, or large genera small species, his calculations would not have supported his theory. His statistical tables thus served to pro-

---

*actuelle*, 2 vols. (Paris: Librairie de Victor Masson, 1855). Darwin's own copy of this book is heavily weighted with annotations.

14. Janet Browne shows that Darwin's statistical analysis had several precedents, most notably in Alexander von Humboldt's *Essai sur la géographie des plantes; accompagné d'un tableau physique des régions équinoxiales* (Paris: Chez Levrault, 1805). Darwin had Humboldt's book with him on the *Beagle*. See Janet Browne, "Darwin's Botanical Arithmetic and the 'Principle of Divergence,' 1852–1858," *Journal of the History of Biology* 13 (1980): 53–89. Darwin's friend Joseph Hooker was quite familiar with different kinds of botanical calculations, and the two corresponded frequently in late 1857 and 1858 about the ratio of species in large genera to those in small genera and about what those ratios meant for his theory.

15. Darwin operationalized "largeness" and "smallness" in this way: count the total number of species in a given flora book and then examine the total number of species in the smallest genera (e.g., say, 10 genera with 1 species each for a total of 10 species); add to that number the total number of species in the next largest genera (e.g., say 15 genera with 2 species each, for a running total of 40 species); keep this up till you reach approximately half the total number of species in the flora book (e.g., say you reach half the total number when you count 50 genera with 4 species each). Then a small genus will be the one holding half the entire number of species but with the fewest species in each genus (e.g., the small genera being those from 1 to 4 species each). A large genus will be those holding the remaining half listed in the book (e.g., those holding 5 species or more).

16. In a splendid essay, Karen Parshall explains Darwin's methods and reanalyzes his statistical conclusions. See Parshall, "Varieties as Incipient Species: Darwin's Numerical Analysis," *Journal of the History of Biology* 15 (1982): 191–214.

17. See Charles Darwin, *Charles Darwin's Natural Selection: Being the Second Part of His Big Species Book Written from 1856 to 1858*, ed. R. C. Stauffer (Cambridge: Cambridge University Press, 1975), 134–67 (hereafter referred to as the *Big Species Book*).

vide, as he wrote his friend Joseph Hooker, "the most important arguments I have met with, that varieties are only small species [i.e., incipient species]—or species only strongly marked varieties."[18]

Darwin's calculations also indicated that the dominant or most common species—those that ranged widely in open areas—were those most conducive to the production of multiple varieties and, ultimately, multiple daughter species. He had three reasons for this suspicion even before doing his calculations, and these reasons, especially the third, reveal hidden aspects of his principle of natural selection. The first reason was simply that in larger areas, there would be more places in the economy of nature for subportions of a common species to fill, that is, to become adapted to.[19] The second reason was that in large areas there would be dynamic interaction and competition among different varieties, different species, and different genera—thus accelerating the adaptive response.[20] Before the 1850s, Darwin had assumed that the selecting environment, that to which animals had to adapt, would be the very slowly changing geological environment: climate, water, and food supply.[21] But he came to realize that it was the proximate and dynamic environment of other species that constantly acted in natural selection. I trace the origin of this new awareness of a dynamic environment later in this chapter.

The third reason Darwin offered for expecting common or dominant species to yield more subspecies is the most telling. Simply, it has to do with the character of large numbers. He believed that larger populations of individuals, accommodated in extensive, open areas, would contain by chance more individuals with favorable variations than would be found in smaller populations. This simple assumption had confirmation in the practice of successful

18. Darwin to Joseph Hooker (1 August 1857), in *Correspondence of Charles Darwin*, 6:438. Darwin's judgment that large genera tended to have large species was based on his "eyeballing" of the ratios. Parshall has shown, in her "Varieties as Incipient Species," that if one runs modern statistical tests on Darwin's ratios, assuming the usual significance levels, the null hypothesis cannot be rejected—that is, one cannot argue that the observed tendencies are the result of something other than simple chance. Parshall notes that Darwin was almost right: if one adopted somewhat larger significance levels than usual, the hypothesis would be acceptable. His eyeball was pretty good.

19. Darwin, *Big Species Book*, 252; Darwin, *Origin of Species*, 102.

20. Darwin, *Origin of Species*, 106: "if some of those many species become modified and improved, others will have to be improved in a corresponding degree or they will be exterminated." There is a comparable passage in the *Big Species Book* (254), but without the sharp, assertive expression of the *Origin*. The *Big Species Book* seems to give more weight to the isolation of groups by geographical barriers (254–61). Darwin also mentions in the *Origin* (104–5) the important role isolation might play in giving varieties a chance to gain a foothold before competition with other species might eliminate them; the balance, however, is yet given to large open areas (105).

21. See, for example, Darwin, "Essay of 1844," 91–93, 156–68.

nurserymen, who raised seedlings in very large numbers; as a consequence they were more apt to discover desired variations than amateur florists who raised only a small number of plants.[22] In the *Origin*, Darwin frequently reiterated that "there will be a better chance of favorable variations from the large number of individuals of the same species" than from a smaller number.[23] It was an elemental matter of mathematical probability. What he did not reckon, however, was that large numbers were effective for the breeder because the latter could search the multitude of individuals for those with desired traits, bring them together, and mate them to produce a new, successful variety. In the wild, the advantageous traits manifested by a few individuals would likely be swamped out when they bred with surrounding individuals having average or unfavorable traits. Darwin had recognized the swamping problem quite early. In the essay of 1842, he wondered if there were anything comparable to the breeder's selection going on in nature: "But is there any means of selecting those offspring which vary in the same manner, crossing them and keeping their offspring separate and thus producing selected races; otherwise as the wild animals freely cross, so must such small heterogeneous varieties be constantly counter balanced and lost, and a uniformity of character preserved."[24]

Nature needed some way to bring individuals with favorable variations together for mating. Larger numbers per se would thus not be more advantageous to the production of distinctive subspecies; without nature having some means of selecting that was comparable to the breeder's intentional selecting and segregating, favorable traits would simply languish and then melt away. Darwin seems to have been misled by the analogy with artificial selection. He simply assumed that natural selection would, like the breeder, resolve the difficulty. (Today, analogous to the problem of swamping is that of gene flow among subpopulations: to allow incipient species—that is, well-marked varieties—gradually to distinguish themselves, gene flow between such groups must remain low.)

Darwin believed that the problem of swamping might be mitigated by what is today called "sympatric speciation"—that is, species production utilizing ecological and behavioral barriers. Originally, though, in the essays of 1842 and 1844, he had maintained that geographical boundaries holding small populations would be optimal for species production; speciation would occur in an *allopatric* way (to use the modern term). Consonant with his new ideas

22. Darwin, *Big Species Book*, 136–37.
23. Darwin, *Origin of Species*, 105. See also 41, 70, 102, 110, 125, 177, and 179.
24. Darwin, "Essay of 1842," 5.

about dominant species and their relation to large genera, however, he now proposed, in the 1850s, that ecological and behavioral barriers alone would be effective in dealing with the swamping problem: "We must not overrate the effects of intercrosses in retarding natural selection; for I can bring a considerable catalogue of facts, showing that within the same area, varieties of the same animal can long remain distinct, from haunting different stations, from breeding at slightly different seasons, or from varieties of the same kind preferring to pair together."[25]

Most biologists today regard sympatric speciation to be a rare occurrence, if it occurs at all. For it to take place, a group would have had initially to achieve reproductive isolation—which in a freely mixing population would be unlikely.[26] In the passage just quoted, Darwin simply presumed the problem to be solved—basically, I believe, because it was solved in artificial selection. He did, however, make a few other assumptions about isolating barriers that softened the difficulties, at least in his own mind; these I consider later in this chapter.

## DIVERGENCE IN THE *BIG SPECIES BOOK* AND IN THE *ORIGIN OF SPECIES*

With the presumptively established facts of his statistical examinations, Darwin then turned, in the *Big Species Book*, to explain exactly how individual organisms diverged from one another to create varieties and how these varieties further diverged to become species. He maintained:

> from the species of larger genera tending to vary most & so to give rise to more species, & from their being somewhat less liable to extinction, I believe that the genera now large in any area, are now generally tending to become still larger. . . . Here in one way comes in the importance of our so-called principle of divergence: as in the long run, more descendants from a common parent will survive, the more widely they become diversified in habits, constitution & structure so as to fill as many places

25. Darwin, *Origin of Species*, 103. The comparable passage occurs in the *Big Species Book*, 257–58.

26. Ernst Mayr was the major proponent of the necessity of geographical isolation to produce what is now called allopatric speciation as opposed to speciation without such barriers, or sympatric speciation. See, for instance, Mayr, "Darwin's Principle of Divergence," *Journal of the History of Biology* 25 (1992): 343–59. His view has become the orthodox position; see, for example, Jerry Coyne and H. Allen Orr, *Speciation* (Sunderland, MA: Sinauer Associates, 2004): "Although the resurgence of interest in sympatric speciation has produced a deluge of new information about ecology, biogeography, and systematics, these data have not supported the view that sympatric speciation is frequent in nature, either overall or in specific groups" (175).

> as possible in the polity of nature, *the extreme varieties & the extreme species will have a better chance of surviving or escaping extinction*, than the intermediate & less modified varieties or species . . . *the principle of divergence always favoring the most extreme forms* & consequently leading to the extinction of the intermediate and less extreme, will taken together give rise to that broken yet connected series of living & extinct organisms, whose affinities we attempt to represent in our natural classifications.[27]

This passage from the *Species Book* expresses four general ideas: (1) as members of a given species spread throughout a large area, they will tend to become more diversified, forming distinct varieties, which themselves, over time, will tend to form distinct species; (2) places—we would say "niches"—exist in nature; (3) the extreme groups—that is, those more diversified from the parent group and other daughter groups—will better be able to fill those places, having the advantage over the intermediate groups, which will thus be subject to greater extinction; and (4) this diversification over time will allow naturalists to classify living and extinct groups into the Linnaean taxonomic categories of variety, species, genus, family, and so on. The second and third ideas are the most problematic. Darwin does not postulate, at least in this passage, that these places in nature are initially unoccupied. He does mention in the *Big Species Book* that "an unoccupied or not perfectly occupied place is an all important element in the action of natural selection."[28] In the *Origin*, he refers to places in the polity of nature that "can be better occupied."[29] Whether there are niches in the economy of nature, occupied or not—that is, whether abstract gaps exist in morphological space independent of the kind of organism, or whether the specific kind of organism creates its own niche—has become an issue principally in late twentieth-century biology.[30] I will, therefore, not pursue this existential question. Pearce has shown in considerable detail that Darwin accepted the antecedent existence of such places in the economy of nature and that he had ample support among other naturalists of the period

27. Darwin, *Big Species Book*, 238 and 273 (my emphases).

28. Ibid., 252.

29. Darwin, *Origin of Species*, 108.

30. The thesis that organisms construct their own niches is most commonly associated with Richard C. Lewontin. See, for instance, R. C. Lewontin, "Organism and Environment," in *Learning, Development and Culture*, ed. E. C. Plotkin, 151–70 (New York: Wiley, 1982); Lewontin, "Gene, Organism, and Environment," in *Evolution from Molecules to Men*, ed. D. S. Bendall, 273–85 (Cambridge: Cambridge University Press, 1983); and Lewontin, *The Triple Helix: Gene, Organism and Environment* (Cambridge: Harvard University Press, 2000).

for this assumption.[31] I believe the third of these ideas—that divergence "favors the extreme forms"—is the most revealing for Darwin's general theory; he proposed it as an explanation for the fact stated in the first idea, that is, that taxonomic groupings are the result of nature favoring extremes. This was indeed a new aspect of his work on divergence. It was not an idea present in his essay of 1844 or in earlier notebooks.

Darwin seems to have conceived the proposal in the mid-1850s that extreme forms had the advantage. In a loose note dated 23 September 1856, he specified a benefit of greater divergence: "The advantage in each group becoming as different as possible, may be compared to [the?] fact that of division of ~~land~~ labour Most people can be supported in each country—Not only do the individuals of each group strive one against the other, but each group itself with all its members some more numerous some less are struggling against all other group[s], as indeed follows from each individual struggling."[32]

The note is vague but seems to argue that (1) there is advantage in varieties and species becoming maximally different from one another; (2) the same kind of advantage occurs in the division of labor (i.e., Milne-Edwards's division of physiological labor);[33] and (3) natural selection acts on this advantage, causing struggle among groups. The precise nature of the advantage is not clear in the note, which is why Darwin may not have initially included the notion in the sixth chapter of the *Big Species Book*.

By March 1857, Darwin had a first draft of his *Big Species Book* chapter on natural selection but with only slight mention of divergence. During the next several months, he added some forty manuscript pages on the principle of divergence, completing these in spring of 1858.[34] Only in these later emendations did he start working out the nature of the advantage—or advantages—divergence was supposed to convey. In addition to the advantage of filling "as many places as possible in the polity of nature," he specified yet another benefit of divergence. In September 1857, he wrote Asa Gray and mentioned this advantage: "One other principle, which may be called the principle of divergence, plays, I believe, an important part in the origin of species. The same

31. Trevor Pearce has worked out the complex history of the concepts "place in nature," "polity of nature," and "economy of nature." See Trevor Pearce, " 'A Great Complication of Circumstances'—Darwin and the Economy of Nature," *Journal of the History of Biology* 43 (2010): 493–528.

32. Charles Darwin, manuscript notes, DAR 205.5.171, Department of Manuscripts, Cambridge University Library.

33. Darwin read Henri Milne-Edwards, *Introduction à la zoologie générale* (Paris: Victor Masson, 1851), in the early 1850s. His copy is extensively marked.

34. See the chronology furnished by the editor, R. C. Stauffer, of the *Big Species Book*, 213.

spot will support more life if occupied by very diverse forms: we see this in the many generic forms in a square yard of turf."[35]

In the added material to the *Big Species Book*, Darwin cited George Sinclair, who showed that a plot of land with only 2 species of grass bore on average 470 plants per square foot, but one with 8 to 20 different species had about 1,000 plants per square foot.[36] Sinclair's experiment supplied Darwin with empirical evidence that divergence produced more abundant life in given locations and a progressive abundance overall. In the *Big Species Book*, he claimed that this empirical result had the sanction of Milne-Edwards's doctrine of the "division of labour"—something Darwin suggested in his note of September 1856. According to Milne-Edwards, creatures having diverse organs fulfilling different functions were higher in the scale of life than those simpler creatures in which different functions were confined to the same organ; for example, those creatures would be "higher" that had a stomach for digestion and lungs for respiration instead of only a stomach that had to perform both functions.[37] Analogously, Darwin claimed, descendants of a carnivore would benefit if some specialized in large prey, others in small prey.[38] This was another case in which the extremes had the advantage.

In the *Big Species Book*, then, Darwin describes two distinct advantages that are supposed to accrue to great divergence: (1) the more extreme groups will be able to occupy more places in the polity of nature; and (2) extreme or divergent groups will ultimately produce more life and, presumably, higher life. In the *Origin of Species*, Darwin economically joins these two advantages in a succinct statement of the principle: "the more diversified the descendants from any one species become in structure, constitution, and habits, by so much will they be better enabled to seize on many and widely diversified places in the polity of nature, and so be enabled to increase in numbers."[39]

But are these really advantages? Why should increased numbers—more life—be an advantage? For whom or what? Why should extreme groups be better able to seize on places in the polity of nature? Why would not interme-

35. Darwin to Asa Gray (5 September 1857), in *Correspondence of Charles Darwin*, 6:448.

36. Darwin, *Big Species Book*, 229.

37. Milne-Edwards, in his *Introduction à la zoologie générale*, mentioned precisely this example in respect to a simple hydra (43–44). Darwin uses the example in the *Big Species Book*, 233. Darwin asserted to Hooker that he thought Milne-Edwards's notion of the division of labor to be the surest criterion for highness or lowness in the scale of life. See Darwin to Joseph Hooker (27 June 1854), in *Correspondence of Charles Darwin*, 5:197.

38. Darwin, *Big Species Book*, 233.

39. Darwin, *Origin of Species*, 112.

diate groups do just as well, or better? And finally: Is divergence—or the production of an extreme form—really a trait that can be selected for?

Some commentators do suggest Darwin held that more life was an advantage and thus a cause of divergence.[40] Darwin himself, though, seems to have regarded it more as a consequence of divergence and not an advantage selected for initially. In a miscellaneous note, dated 30 June 1855, he compared two different environments, one thick with heather and the other a variegated meadow. He deemed the latter conducive to the production of more life and concluded, albeit with hesitation: "This is not final cause, but more result from struggle (I must think out this last proposition)."[41] This seems the logically appropriate judgment, namely, that more life was a consequence instead of a cause. However, the other two questions linger: Why should extreme forms have the advantage, and is great divergence really a trait that can be selected? To get another perspective on Darwin's treatment of divergence and these questions, I turn to some of the scholarly literature on the subject, a literature that is extensive and itself divergent in its interpretations.[42]

## SCHOLARLY INTERPRETATIONS OF DARWIN'S PRINCIPLE OF DIVERGENCE

I briefly examine three representative interpretations of Darwin's principle of divergence, since they indicate some of the perplexities of his account. In

40. Schweber cites the just-quoted passage from the *Origin* and suggests that the two advantages of divergence are securing a place in the polity of nature and producing more life. See Silvan S. Schweber, "Darwin and the Political Economists: Divergence of Character," *Journal of the History of Biology* 13 (1980): 195–289. He sums it up this way: "Adaptation toward a place in the economy of nature together with the principle of the maximum amount of life per unit area as the overall driving force make understandable why there is divergence of character: in ecological differentiation and adaptation the primary factor of divergence is functional specialization" (212).

41. Darwin, manuscript notes, DAR 205.3.167.

42. See, for example, Dov Ospovat, *The Development of Darwin's Theory: Natural History, Natural Theology, and Natural Selection, 1838–1859* (Cambridge: Cambridge University Press, 1981), 170–90; Schweber "Darwin and the Political Economists"; Browne, "Darwin's Botanical Arithmetic and the 'Principle of Divergence'"; Frank Sulloway, "Darwin and His Finches: The Evolution of a Legend," *Journal of the History of Biology* 15 (1982): 1–53; David Kohn, "Darwin's Principle of Divergence as Internal Dialogue," in *The Darwinian Heritage*, ed. David Kohn (Princeton: Princeton University Press, 1985), 245–57; Barbara Beddall, "Darwin and Divergence: The Wallace Connection," *Journal of the History of Biology* 21 (1988): 1–68; Mayr, "Darwin's Principle of Divergence"; William Tammone, "Competition, the Division of Labor, and Darwin's Principle of Divergence," *Journal of the History of Biology* 28 (1995): 109–31; David Kohn, "Darwin's Keystone: The Principle of Divergence," in *The Cambridge Companion to the "Origin of Species,"* ed. Michael Ruse and Robert J. Richards (Cambridge: Cambridge University Press, 2009), 87–108; and Pearce, "'A Great Complication of Circumstances'—Darwin and the Economy of Nature."

1992 Ernst Mayr focused on Darwin's letter to Asa Gray, which he believed encapsulated the principle and its rationale. Mayr wrote: "The basic point of the principle of divergence is simplicity itself: the more the coinhabitants of an area differ from each other in their ecological requirements, the less they will compete with each other; therefore natural selection will tend to favor any variation toward greater divergence. The reason for the principle's importance to Darwin is that it seemed to shed some light on the greatest of his puzzles—the nature and origin of variation and of speciation."[43]

So for Mayr, Darwinian divergence is a trait (1) favored by selection; (2) favored because it reduces competition; and (3) believed by Darwin to explain the production of varieties and species. In the bulk of his essay, Mayr disputed this last point, arguing that Darwin really could not adequately explain speciation. The splitting of species required, in Mayr's estimation, geographical isolation, whereas Darwin thought speciation would occur more readily in large, open areas—thus "sympatric speciation." As I indicated earlier, Darwin had initially assumed that geographical barriers were necessary for the production of new species.[44] And in the *Origin of Species*, he did note some of the facilitating features of geographical isolation, for instance, on islands.[45] But during the 1850s, he came to hold that large open areas were more conducive to the production of species, and this is the general position maintained in the *Big Species Book* and in the *Origin of Species*. I will discuss the role of the environment in more detail later in this chapter.

While Mayr and others believe that Darwin allotted the advantage of divergence to reduction in competition, William Tammone contends that Darwin never claimed divergence to have this advantage.[46] Tammone points out that Darwin usually spoke of species coming into already *occupied* places in nature and therefore that such places would be subject to ongoing competitive struggle.[47] The advantage of divergence for Darwin, according to Tammone, is that it produced greater specialization: "But if the advantage of divergence is not reduced competition, then what is it? As I have already suggested, the so-called advantage of divergence is that it leads to increased specialization. This is because increased specialization makes an organism more skill-

43. Mayr, "Darwin's Principle of Divergence," 344.

44. Darwin, "Essay of 1844," 91–93, 156–68.

45. Darwin, *Origin of Species*, 104–5.

46. Actually Darwin made precisely that claim in his loose note of 23 September 1856, quoted in the previous section; he did not, however, reiterate it in the *Big Species Book* or in the *Origin of Species*.

47. Tammone, "Competition, the Division of Labor, and Darwin's Principle of Divergence," esp. 118–19.

ful or more competent in securing the resources necessary for survival and reproduction."[48]

Tammone stresses the analogy that Darwin drew with Milne-Edwards's principle of the division of labor, which describes the benefits of specialization of parts internal to a biological organism; the comparable advantage would go to lineages that diverged for greater specialization. He also indicates that for Darwin, divergence not only led to greater specialization but to competitive exclusion of closely related organisms—the parent species, Darwin supposed, is usually driven to extinction since the daughter species "improve" on it.[49] Because of both divergence and extinction, gaps would be produced among species and thus would allow for the application of the Linnaean categories.

Mayr and Tammone agree that divergence is a trait that is favored, though they disagree about why it is favored: for Mayr, because it excludes competition; and for Tammone, because it leads to increased competition, yielding greater specialization and a better hold on a place in nature. Mayr and those agreeing with him (e.g., Frank Sulloway and Chris Haufe) seem to have put specialization and competition in the wrong order.[50] Darwin certainly maintained that the extreme forms—those more divergent from parent and sibling forms—would have the advantage in securing a place in the polity of nature. If it is a different place than that occupied by similar forms, then less competition would be the result; if it is virtually the same place, then less competition would also result since the previous occupant would ultimately be forced to vacate its place and, perhaps, be driven to extinction. In both instances, less competition would be a *consequence* of specialization; it would not be the initiating advantage. So Tammone seems correct in his assessment. What he has neglected, however, are the two questions I posed earlier: Why should forms that are extreme have the advantage? And is extreme divergence a trait that can be selected for? Another scholar who has written on Darwin's principle of divergence, David Kohn, highlights these issues.

Kohn contributed an essay on divergence to the *Cambridge Companion to the Origin of Species*, which Michael Ruse and I edited (2009). In that essay,

48. Ibid., 122.

49. Darwin, *Origin of Species*, 119, 128.

50. See Frank Sulloway, "Geological Isolation in Darwin's Thinking: The Vicissitudes of a Crucial Idea," *Studies in History of Biology* 3 (1979): 23–65. In a personal communication, Chris Haufe put it to me this way: "The less dependent an individual is upon the set of resources towards which the rest of the population is oriented, the less pressure there is on that individual to outcompete other members of the population for that set of resources."

Kohn argued that Darwin's principle of divergence involved what he called "divergence selection," a kind of natural selection that picked out the extremes or most divergent forms: "When Darwin deployed the principle of divergence, he always did so in conjunction with natural selection. The principle acts as an amplifier of selection. This coupling of divergence and selection created a special case or type of natural selection, which we may term divergence selection. This is selection where conditions favor divergent specializations among related forms sharing a common location." Kohn points out that with abundant variation, there would be different forms available to exploit different features of an expansive environment. And so "this situation will favor selection of the most extreme—that is, the most divergent—forms."[51]

As I edited Kohn's draft, I questioned this formulation. I put it to him: "Isn't all selection divergent selection?" This is because all selection picks out individuals with slightly different traits, favoring some and excluding others such that morphological gaps would result. Extreme forms would then be a *consequence* of ordinary selection over long periods of time.[52] Kohn, however, strongly dissented. He responded: "Here we disagree. No, not all selection leads to divergence or 'is divergent.' You can't mean what I think is the plain meaning of your statement. Of course all selection leads to being different from an ancestor, but divergence means more than mere difference and/or deviation from ancestors. Rather it means the multiplication of lineages in different directions. That at least is the problem CD is trying to solve in this part of the *Origin*: namely, the problem of explaining branching by means of natural selection."[53]

A scholar of David Kohn's talents gets the last word on his own essay, and so his original formulation stands in the Cambridge volume. For Kohn, the principle of divergence is an amplification of selection or a kind of natural selection in which extreme forms have the advantage and are thus selected. Elliott Sober has come to share Kohn's interpretation.[54]

Kohn's interpretation still seems incoherent to me, or at least inconsistent with the major thrust of Darwin's theory. One needs to consider natural selection on the ground, as it were. When a parent form produces several offspring, they presumably differ only slightly from one another and from the parent form itself, with one or another of the progeny having a small advantage in a given

51. Kohn, "Darwin's Keystone," 88, 91.

52. Robert J. Richards to David Kohn, pers. comm., June 2007.

53. David Kohn to Robert J. Richards, pers. comm., July 2007.

54. Elliott Sober, *Did Darwin Write the Origin Backwards? Philosophical Essays on Darwin's Theory* (Amherst, NY: Prometheus Books, 2011), 33–34.

environment. From moment to moment, selection, by whatever name, can only choose just those small, individual differences that provide the competitive edge. It cannot choose the extreme form, except that the extreme form just happens to fit in a given environment. There is, however, no reason why such a fortuitous fit should be antecedently expected—indeed, extreme slowness, extreme size, extreme color would more likely be extremely detrimental in the struggle for life. This interpretation verges on the assumption that divergence selection requires "hopeful monsters." But if a hopeful monster were produced here or there, it would likely be swamped out by others having only average or even negative traits, since such outliers must be rare. Moreover, natural selection does not simply work on one trait but on all of the traits that an animal exhibits—some perhaps of an advantage, others of a disadvantage. Consequently any one generation of animals will have a fine gradation of more or less successful organisms that will promiscuously mate, with the consequence that the population will, under the most favorable circumstances, only very gradually change. Swamping out continued to be a problem for Darwin.

Consider this scenario about wild dogs in a given location in Australia. Some, by chance, are slim and quite fast; others, not so fast, but with slightly more muscular bodies and bigger paws. Both groups compete for rabbits, with the former slowly improving their speed from generation to generation. If the latter begin to discover a mole here or there, and these more clumsy animals begin to compete with one another in the digging for moles in the hard, encrusted ground, though still occasionally running down slower rabbits, then the original groups may begin to diverge, with individuals of each, however, continuing to compete within their respective groups. For all individuals, though, selection would be choosing not extreme traits but traits that by chance would give a slight competitive advantage in a particular habitat. As the two groups further diverge and the individuals within each group increase the competitive ante, new varieties would slowly be formed. Extreme forms may gradually emerge, but not because selection is picking out extreme forms; in all instances, selection would be acting on just slight differences among close competitors. Divergence in this scenario would thus be a long-term consequence of ordinary selection, not a special kind of selection. In essence, this was Darwin's position in the 1844 essay. The answer to the questions I asked about whether extreme divergence was an advantage and a trait that could be selected for must be no. No postulation of a special principle of divergence was, therefore, necessary.

Divergence selection, as Kohn proposed it, could only occur if selection could see into the future and select the series of extreme differences that would have an ultimate goal, namely, some greatly divergent form. It's not too much of an exaggeration to say that Kohn is postulating a hopeful monster as the kind of variation that divergence selection would be working on. Yet he may well be truer to Darwin's new conception of the 1850s than my own counterclaim supposed. To see this, we need to look at the model Darwin newly introduced in both the *Big Species Book* and the *Origin* to explain the operation of the principle of divergence.

## DARWIN'S MODEL OF DIVERGENCE

Let me quote again Darwin's principle of divergence as it appears in the *Origin*:

> the more diversified the descendants from any one species become in structure, constitution, and habits, by so much will they be better enabled to seize on many and widely diversified places in the polity of nature, and so be enabled to increase in numbers.[55]

Just before he offered this definition, Darwin asked his reader to consider the practice of domestic breeders.

> A fancier is struck by a pigeon having a slightly shorter beak; another fancier is struck by a pigeon having a rather longer beak; and on the acknowledged principle that "fanciers do not and will not admire a medium standard, but like extremes," they go on . . . choosing and breeding from birds with longer and longer beaks, or with shorter and shorter beaks. . . . Here, then, we see in man's productions the action of what may be called the principle of divergence, causing differences, at first barely appreciable, steadily to increase, and the breeds to diverge in character both from each other and from their common parent.[56]

The breeder thus selects the most extreme traits and ultimately winds up with a morphologically very extreme individual. Darwin believed nature acted analogously: she chooses extreme traits *at every iteration* and finally produces a quite distinct species. The advantage realized would be a more secure hold on resources and greater numbers. Darwin's emphasis on the principle of di-

55. Darwin, *Origin of Species*, 112.
56. Ibid.; the comparable passage in the *Big Species Book* is at 227–28.

vergence as "favoring extremes" drew inspiration from the practice of breeders who also favored extremes.

Darwin's appeal to artificial selection as a model for processes in nature certainly conforms to his general strategy in the *Origin of Species*, but I believe it had a special initiating cause in this instance. In spring 1855, shortly after he had begun work on the *Big Species Book*, he decided that he needed experience in the breeder's art. His initial motivation for undertaking this messy practice, as he explained to his cousin William Darwin Fox, was to determine when the very young of related breeds began to show characteristic differences.[57] He had been convinced from his earliest years that organisms would repeat in their ontogenetic development the morphological patterns of their ancestor species, and now he would conduct exact measurements to reveal the transmutational past of domestic animals.[58] He had been persuaded by William Yarrow, an experienced breeder, to try pigeons for this purpose.[59] His first effort was to observe when the distinctive feathers of the fantail pigeon would appear in ontogenesis and begin to distinguish the fantail from other breeds. Darwin started this enterprise with hesitation but soon felt real enthusiasm for the pigeon fancier's art.[60] He had breeding stalls built in his back garden and joined two popular pigeon-breeding clubs. He carried on many breeding and dissection experiments up through 1858. The focus of his effort was to demonstrate that the wildly divergent pigeon breeds had all derived through domestic selection from the common rock pigeon, *Columba livia*. The hatchlings and squabs of different domestic breeds recapitulated their earlier forms and thus gave evidence of a common ancestor. Demonstration of this descent provides a central aim of the first chapter of the *Origin*.

Darwin read several books on pigeon breeding, especially the treatises by John Eaton. In his *Variation of Animals and Plants under Domestication*, Darwin quoted Eaton's dictum: "Fanciers do not and will not admire a medium standard, that is, half and half, which is neither here nor there, but admire extremes."[61] Eaton's remark is echoed in the passage from the *Origin* about domestic selection, which I quoted at the beginning of this section. It appears,

57. Darwin to W. D. Fox (19 March 1855), in *Correspondence of Charles Darwin*, 5:288.

58. I have traced Darwin's deployment of the principle of recapitulation from his early notebooks to the last editions of the *Origin of Species*. See Richards, *Meaning of Evolution*, chap. 5.

59. Darwin to W. D. Fox (27 March 1855), in *Correspondence of Charles Darwin*, 294.

60. James Secord provides a full account of Darwin's efforts at pigeon breeding in Secord, "Charles Darwin and the Breeding of Pigeons," *Isis* 72 (1981): 162–86.

61. John M. Eaton, *A Treatise on the Art of Breeding and Managing Tame, Domesticated, Foreign and Fancy Pigeons* (London: printed by the author, 1858), 86. See Darwin's discussion, *Variation of Animals and Plants under Domestication*, 2 vols. (London: Murray, 1868), 1:215.

then, that Darwin's conception that nature favored extremes came from his experience with breeding pigeons during the period when he was formulating his principle of divergence. Pigeon fanciers went after extremes, and, he assumed, nature did as well.

This answers the question I put earlier: What did Darwin think he had missed in the 1844 essay and what element was new to his consideration of the problem of morphological divergence in the 1850s? What must have struck him during his carriage ride was the practice of breeders in producing wildly divergent races of pigeons. What seems to have escaped his reflective notice, however, was the salient difference between nature and the breeder: the pigeon fancier can detect extreme traits and carefully select out of his flock just those birds that display such traits, segregate them, and mate the individuals together. Nature, it would seem, cannot accomplish a comparable feat. I believe Darwin, nonetheless, became convinced that the analogy with artificial selection was apt because of four other assumptions he made: the dynamism of the environment; keener competition in large open areas; greater extinction in intermediate zones between stations; and natural selection as an intentional agent. I will discuss the first three in the next section and the last thereafter.

## DARWIN'S CHANGING ASSUMPTIONS ABOUT THE ENVIRONMENT

The environment plays three general roles in Darwin's developing theory: (1) it produces variations in animals by acting on the reproductive organs of parents and in Lamarckian fashion by direct impact or by stimulating animals to adopt new, altering habits; (2) it segregates organisms from each, so that swamping is prevented; and (3) it acts as a selecting agent, operating on favorable variations over time.

In his early notebooks, Darwin assumed that adaptive variations in organisms would be directly induced by the actions of the environment and that these alterations would be heritable. He thought sexual generation made new offspring particularly susceptible to such alteration. He quickly came to understand, however, that direct environmental impact was too crude to produce intricate adaptations. He then supposed that changes in the environment also might induce animals to adopt new habits, which would more finely adapt them to their circumstances, and that these habits, through practice over generations, would become innate instincts; finally, these instincts would gradually alter anatomy. After he formulated the rudiments of his device of natural selection, he continued to assume that the environment would directly alter organisms and thus supply the variations upon which natural selection would work. Only in the 1842 essay did he maintain that a principle source of variation available to selection was the action of the environment on the reproductive organs of the parents.[62] But whether variations were directly induced or indirectly through the parents, Darwin simply presumed that they would occur infrequently in the wild; this was because, unlike the domestic situation, environmental change in nature occurred with Lyellian slowness.

In his early notebooks, Darwin maintained that isolation of a group of animals or plants—for example, on an island—would gradually alter the group's character to form a new species. He initially supposed this to have been the case with mockingbirds blown over to the Galapagos Islands from the mainland. They wound up on different islands, and the pressures of the local environments altered, in a Lamarckian fashion, their morphological structure sufficiently for them to be regarded as distinct species. Even after he formulated his device of natural selection, he continued to argue, in light of artificial selection, that physical isolation was a principal factor in the formation of new species. After all, the successful breeder would segregate just those animals with the desired traits for mating, thus keeping the traits from being swamped out by backcrosses to unfavored individuals. Geographical barriers would serve the analogous function of the breeder in preventing promising variations from being dissipated, something Darwin affirmed in the essay of 1844: "isolation as perfect as possible of such selected varieties; that is, the preventing their crossing with other forms; this latter condition applies to all terrestrial animals, to most if not all plants and perhaps even to most (or all) aquatic organisms."[63] He conceived two distinct possibilities for the isolation necessary to create new species. Either animals or plants would have settled on islands, like the Galapagos, and there become adapted by selection to their circumstances, or portions of a continent would subside, with the higher areas forming islands on which animals and plants would be isolated. These organisms would undergo adaptation, and then with uplift, what had been separate stations would be reconnected. Thus, new species would have been formed while geographically segregated, and their reproductive isolation would keep them distinct after connections had been reestablished.[64] Since the newly formed proto-species would be tightly adapted to their habitats, the

62. Darwin, "Essay of 1842," 2.
63. Darwin, "Essay of 1844," 183.
64. Ibid., 189–90.

intermediate corridors now connecting the formerly isolated areas would be inhospitable to the new groups; any migrants attempting to colonize the intermediate zones would be few in number and ill equipped to adapt to those connecting areas. Intermediate groups would thus be susceptible to extinction: because of fewer numbers, their chance of survival would be less; and because of greater competition along the periphery from the extremes, they would be more easily extirpated. In this scenario, the extremes would be preserved and the intermediates extinguished—hence the gaps between species. Darwin would retain the notion of the disadvantage of groups in the intermediate zones when he came to see the potency of ecological barriers; they would function like geographical barriers.

Darwin was a conservative thinker. Ideas that he once formulated, he tended to retain in his later theorizing, even if they had to undergo modifications. His views about the function of geological barriers became subordinated to his new conception, in the 1850s, about the formation of species in large open areas, but he never relinquished the notion that in some instances species were produced very slowly through the isolating mechanisms of geological change.[65] This retention led to some strikingly contradictory assertions in the text of the *Origin*. So in some places, not abandoning his belief in slow geological change as a source of variation, he suggests that favorable variations might arise in a species only "in the course of thousands of generations" and that, as a result, natural selection would operate only infrequently over very long periods: "I do believe that natural selection will always act very slowly, often only at long intervals of time and generally on only a very few of the inhabitants of the same region at the same time. I further believe that this very slow intermittent action of natural selection accords perfectly with what geology tells us of the rate and manner at which the inhabitants of this world have changed."[66]

This assumption is in stark contrast to the dominant view in the *Origin*, namely, that "natural selection is daily and hourly scrutinizing throughout the world, every variation, even the slightest; rejecting that which is bad, adding up all that is good."[67] In some instances, then, the text is vague—really, contradictory—about whether natural selection is supposed to be always or only

65. In the *Origin* (107–8), Darwin retained the presumption that continental areas would be subject to subsidence and later uplift, thus providing geological barriers to foster the formation of new species. But the importance of these geological movements became subordinated to the idea of speciation in open areas.

66. Ibid., 81–82 (quote), 83, 108 (quote); see also 80 and 84. Darwin makes comparable remarks in the *Big Species Book*, 261–62.

67. Darwin, *Origin of Species*, 84.

occasionally operating. Wallace noted this conundrum in the *Origin* and recommended Darwin drop the phrases about the intermittent and infrequent action of natural selection. Wallace judged it was truer to his friend's theory to conclude that "variations of every kind are always occurring in every part of every species."[68]

Wallace's admonition was, in a sense, superfluous. In other parts of the *Origin*, Darwin had asserted that systematists recognized that variation from parent to offspring was constant, both in unimportant, peripheral parts and in important, structural parts.[69] He had earlier come to this realization in his work on barnacles in the late 1840s and early 1850s—at least this is what he mentioned to Hooker as one of the consequences of his systematic work for his still nascent theory.[70] Yet something more was needed than frequently available variation for Darwin to perceive that natural selection was always at work.

The supposition that natural selection was constantly acting derived from Darwin's new conviction, reached in the 1850s, that the operative selector in a given environment was not so much the geological features and climate of an area but "the presence of other competing forms better adapted to such conditions." He came to hold that "all nature [was] bound together in an inextricable net-work of relations."[71] This web of life would both constantly vibrate with competing forms and simultaneously create the isolating barriers he had earlier postulated. He thought of these new kinds of barriers as comparable to geographical boundaries: they would form stations in an extended area, with intermediate zones between them. Darwin simply assumed that those intermediate zones, as in the case of geological isolation, would generally be inhospitable to migrants and thus extinction would be fostered. Hence, he presumed that the swamping problem would be mitigated and sympatric speciation effective. But whence his new conception of the web of life?

In the *Big Species Book* and in the *Origin of Species*, Darwin vividly epitomized the intricate relations of creatures with his example of the way more cats

68. Alfred Russel Wallace to Darwin (2 July 1866), in *Correspondence of Charles Darwin*, 14:229.

69. Darwin, *Origin of Species*, 45.

70. Darwin to Joseph Hooker (13 June 1850), in *Correspondence of Charles Darwin*, 4:344. Darwin told Hooker that his barnacle work did not contribute very much to his theory, though he was "struck (& probably unfairly from the class) with the variability of every part in some slight degree of every species when the same organ is *rigorously* compared in many individuals" (344). Browne believes the barnacle work brought Darwin to realize that variation was constant and in all parts of organisms. See Janet Browne, *Charles Darwin Voyaging* (Princeton: Princeton University Press, 1995), 512–15. Darwin, in his conclusion here, does not seem quite confident. In any case, he did not immediately draw the conclusion that if variation was constant, selection should also be constant.

71. Darwin, *Big Species Book*, 266–67.

in a neighborhood would cause clover to become more plentiful: cats would control the field mice that destroyed the nests of humble bees that pollinated the clover.[72] He drew this example from a fleeting passage on humble bees in an entomological journal that he read in late summer of 1854.[73] He had been following the activities of humble bees, which had nests in his back gardens, and in that connection he read the article, which seems to have made him more reflectively aware of the web of organic interaction.[74] In the *Origin*, he immediately followed this example of the humble bees with mention of the "entangled bank" of life, a famous image that forcefully reemerges in the last paragraph of the book.[75] The tangle of life furnished Darwin with a different kind of environment—a dynamic environment. Whereas, in the earlier essays, he relied on very slow geological processes to furnish the selecting environment,[76] he now conceived of that environment as always active, sometimes in a stable tension of finely balanced forms, sometimes in a shifting disequilibrium of rapidly altering forms. A dynamic environment established for him the analogical foundation for his controlling metaphor of natural selection as "daily and hourly scrutinizing, throughout the world, every variation."

Let me take stock of the conclusions of this chapter drawn thus far. Although Darwin had the rudiments of his principle of divergence already in the 1844 essay—that is, his conviction that adaptation to different places in the natural polity would begin to divide incipient varieties of a species—he later came to assume another factor was operative, namely, that nature selected extremes, just as the pigeon fancier did. In a dynamic environment, which he also came to appreciate in the 1850s, the process of natural selection would be ongoing, always selecting extremes. Intermediate individuals, he believed, would be at a disadvantage, hence both preventing extremes from being swamped out and producing gaps among varieties through extinction. That latter conviction stemmed from his early analysis of the role of geological boundaries in producing distinct species, what we would call *allopatric spe-*

72. Ibid., 183; Darwin, *Origin of Species*, 74.

73. An abstract of a paper ("Habits of Bombinatrices") by H. W. Newman was given in *Transactions of the Entomological Society of London*, n.s., 1 (1850–51): 86–94 (section on proceedings); the mention of cats preying on mice that destroyed the nests of humble bees occurs on 88; Darwin added the further relation to clover.

74. Darwin kept a small notebook on humble bees (DAR 194.1–12, Department of Manuscripts, Cambridge University Library) from 8 September to 2 October 1854; he noted the précis of Newman's paper (see previous note) on p. 10 of the notebook. Darwin collected other examples of interaction among organisms; see Darwin, *Big Species Book*, 180–86.

75. The image of an entangled bank does not appear in the *Big Species Book*.

76. Darwin, "Essay of 1844," 83–84.

*ciation*. He now adopted the idea that selection would more readily occur in an open, extensive environment where competition would be keener—that is, under *sympatric speciation*.

Thus far we are approximately at the position Darwin took in the 1844 essay, except instead of geographical barriers, he now supposed ecological barriers, and instead of the intermittent activity of natural selection, he now supposed a constant activity. These features, though necessary assumptions for his principle of divergence to work, do not seem to have that decisive character implied by his eureka discovery while riding in his carriage in the mid-1850s. Was his discovery simply that he recognized that most divergent varieties would have a better chance of seizing on an unoccupied place in the polity of nature—the view proposed by Mayr, Ospovat, and others?[77] Again, as Tammone argued, that seems unlikely. Moreover, there would be no reason, except fortuitous chance, that an extreme form would happen to meet the requirements of an unoccupied niche, much less one that is already occupied. Something more must have moved Darwin decisively. As I've indicated, I think that more was the analogy with the pigeon fancier's selecting extremes. Yet, to make the analogy work—that nature, too, selected extremes—Darwin had to assume a feature of natural selection that clearly displays its nineteenth-century origin. Before I explore in more depth Darwin's image of the operations of natural selection, let me give a brief account of Jerry Fodor's assault on the principle.

## FODOR'S REJECTION OF NATURAL SELECTION IN NEO-DARWINISM

Fodor argues that neo-Darwinian theory fails because it relies on the principle of natural selection, which is fatally flawed: the principle assumes that nature acts from intentions. In their book *What Darwin Got Wrong*, Fodor and Piatelli-Palmarini maintain that recent biological research and theory deploy other mechanisms that can account for evolution without appeal to natural selection. The crux of their argument against natural selection—really Fodor's argument—can be briefly laid out. They assert that any trait assumed to have been selected for has other linked traits that come along with it—"free riders"; for nature to select only one of the linked traits is to assume that nature can discriminate, can form intentions to choose one and not the other, which, of course, it cannot do. When the dog breeder selects, for example, German

77. See Ospovat, *Development of Darwin's Theory*, 176. Haufe also urged this argument to me in a personal note.

shepherds for a certain coat color and skull shape, he or she unintentionally also selects for hip dysplasia (which shepherds notoriously suffer from). The breeder's intention, however, is clear: selection for the one set of traits and not the other. But nature cannot make comparable discriminations. To use Fodor and Piatelli-Palmarini's mildly ludicrous example: What justifies the claim that nature has selected hearts to pump blood and not for hearts to make pumping sounds, a necessarily linked trait? The authors claim that such attribution can only be justified by assuming nature has intentions—she intends to select only for pumping ability; however, since nature does not have intentions and supposed selected traits will always have free riders, the appeal to natural selection can never be justified as an explanation for any trait.

*What Darwin Got Wrong* is a mess. To point out just one general and fatal feature: in their screed against neo-Darwinism, the authors claim that the most recent biological research replaces natural selection with endogenous mechanisms that impose constraints on the development of traits. So Dumbo, the baby elephant, will never fly because his ears would have to be extremely large, though no internal cartilaginous structures could support ears of the required size. The *constraint on* ear size thus determines species characteristics. What the authors fail to recognize, however, is that "constraint on" implies intentions no less than does "natural selection for." To claim that an organism is constrained in a particular way is to assume that it would not be so restricted if a counterfactual situation obtained (e.g., that cartilaginous structures of elephants could support great weight). Yet as the authors note, only intentional systems can be sensitive to contrary-to-fact conditions. Following their logic, therefore, the application of "constraint on" implicitly ascribes intentions to nature. Thus their supposed substitute principle is epistemically no different from natural selection. There are many other problems with their claims, but let me turn to the central argument against natural selection.[78]

When contemporary neo-Darwinists explain some trait by natural selection or by endogenous constraints, they make no implicit assumptions about nature having intentions. Quite routinely, for example, medical experts attribute the evolution of drug-resistant strains of bacteria to the excessive use of antibiotics in hospitals. Scientists understand quite well how selection operates in these instances; indeed, they are able experimentally to breed drug-resistant bacteria precisely in the way these organisms are selected for in the "wild," thereby confirming the natural selection of drug resistance. Appeal to natural selection does involve intentions, but those of the biologist making the ascrip-

78. I have described many other problems with their argument in Richards, "Darwin Tried and True."

tion. He or she judges, on the basis of sustained observation or experiment, that a particular environmental condition is causally sufficient to produce the trait at issue; the judgment is intentional, but it is the biologist's intention, not nature's. And often, as in the case of drug resistance, experiment can demonstrate the causally sufficient conditions for the trait beyond a reasonable doubt. If some apparently linked trait might be a candidate as the cause of survival, the biologist can change the experimental condition to exclude that trait, or can make observations of conditions in which the suspect trait is absent. No explanation based on experiment or observation is immune from the possibility of an alternative explanation, but that is merely the character of all science. Essentially Fodor's screed is a rejection of modern science altogether.

There is no evidence that either Fodor or Piatelli-Palmarini ever read Darwin's *Origin of Species*. Their arguments were directed only to neo-Darwinian biologists. But could they be right in respect to Darwin himself? Did he assume nature had intentions?

## DARWIN'S PRINCIPLE OF NATURAL SELECTION

Darwin's essays of 1842 and 1844 were his first efforts at a systematic formulation of the theory that he had begun to construct in his several transmutation notebooks, beginning in 1837. In those later essays, still feeling his way toward a coherent and encompassing conception, he set out to explain to himself the operations of natural selection. He initially considered how the human breeder transformed his domestic creatures through selection. In that light, he constructed a model of natural selection as a very powerful intelligence that could choose creatures. "With time enough," he thought, "*such a Being might rationally (without some unknown law opposed him) aim at almost any result.*"[79] (See the previous chapter for the full passage.)

This passage from the 1844 essay mirrors a comparable one in the 1842 essay and advances virtually the same model found in the *Big Species Book* and in the *Origin of Species.* I've quoted the relevant passage from the *Origin* in the previous chapter, but let me again cite it, since it is the culmination of Darwin's understanding of the actions of natural selection:

> Man can act only on external and visible characters: nature cares nothing for appearances, except in so far as they may be useful to any being. She can act on every internal organ, on every shade of constitutional

79. Darwin, "Essay of 1844," 85–86 (emphases mine).

difference, on the whole machinery of life. Man selects only for his own good; *Nature only for that of the being which she tends*. . . . Can we wonder, then, that nature's productions should be far "truer" in character than man's productions; that they should . . . *plainly bear the stamp of far higher workmanship*? It may be said that natural selection is daily hourly scrutinizing, throughout the world, every variation, even the slightest; rejecting that which is bad preserving and adding up all that is good; silently and insensibly working, whenever and wherever opportunity offers, *at the improvement of each organic being* in relation to its organic and inorganic conditions of life.[80]

Several features of Darwin's model for natural selection need to be emphasized (and I have done so through the use of italics in the previous two passages), since they explain other aspects of his conception of the principle of divergence.[81] First, the model is that of a rational and moral selector, not a machine. The passage just quoted attributes to natural selection a power of "discrimination" keener than any machine of the period could demonstrate. That discriminatory power might yield a very slow, gradual change in the tree of life, quite different from the rapid, saltational, and mechanistic alterations that Darwin's friend Huxley thought more realistic.[82] The "rational" features of natural selection could thus produce a "far higher workmanship" than even human intelligence might attempt.

The attribution of intelligence to natural selection, at least implicitly, explains certain features of Darwin's conception of the principle of divergence. The swamping problem attendant on the assumption of large numbers of a species in an extended, open area could be overcome if natural selection somehow acted with the intelligence of the breeder, who segregated favorable variations for mating. But even more significantly, a rational selector could select "extremes," thus producing the morphological gaps separating species, genera, and the higher taxa from each other.

80. Darwin, *Origin of Species*, 83–84 (emphases mine).

81. I have considered Darwin's principle of natural selection in chapter 2 of this volume, "Darwin's Theory of Natural Selection and Its Moral Purpose."

82. Thomas Henry Huxley, "Darwin on the Origin of Species," *Westminster Review*, n.s., 17 (1860): 541–70. Huxley lodged this singular criticism: "And Mr. Darwin's position might, we think, have been even stronger than it is if he had not embarrassed himself with the aphorism, 'Natura non facit saltum,' which turns up so often in his pages. We believe, as we have said above, that Nature does make jumps now and then, and a recognition of the fact is of no small importance in disposing of many minor objections to the doctrine of transmutation" (569). Huxley seems not to have noticed that Darwin's principle of divergence does countenance jumps in selection when it operates on extremes.

At the end of the section titled "Scholarly Interpretations of Darwin's Principle of Divergence," I attempted to show that natural selection could eventually produce extremes—that is, diverging species—if it continued to act on small, minute differences in a changing environment, but that it could not iteratively select extremes at each moment after the manner of the pigeon fancier. Or rather, it could select such extremes if it acted rationally and with a goal, just as the pigeon fancier did. In short, the principle of divergence required natural selection to operate in a rational way to achieve the desired end of separating the taxonomic groupings.

A second important feature of Darwin's principle of natural selection, as determined by the model underlying it, is that selection has a moral purview. As emphasized in the just-quoted passage from the *Origin*, natural selection works only for the good of each being which she tends; she works for "the improvement of each organic being." Darwin repeats phrases like these some five times in the *Origin*; so, for example: "And as natural selection works solely by and for the good of each being, all corporeal and mental endowments will tend to progress towards perfection."[83] From our contemporary, neo-Darwinian perspective these expressions are simply in direct contradiction to the logic of natural selection: natural selection does not work for the good of most beings; it destroys most creatures; it eliminates them and their seed. Darwin, however, was so wedded to the model of natural selection as a benevolent, intelligent force that he ignored what we would regard as the very logic of this natural process.[84] (I have examined the moral character of natural selection in greater detail in chapter 2 of this volume.)

## CONCLUSION

Darwin considered his principle of divergence a linchpin for his entire theory. The principle was designed to explain the clustering of organisms into varieties, species, genera, and the higher taxonomic categories. The history of the principle is perplexing. Darwin claimed he only came to see the problem of divergence and its solution in the 1850s, though he seems to have recognized

83. Darwin, *Origin of Species*, 489. Other instances of similar expressions occur at 83, 84, 149, 194, and 201.

84. It is possible that when Darwin wrote that natural selection worked for the benefit of each being, he implicitly took a retrospective view—that is, each creature had realized the advantages obtained by its ancestors. Yet Darwin gave no hint in the text that he assumed that kind of vantage; the remark that as a result of natural selection "all corporeal and mental endowments will tend to progress toward perfection" seems to look at the consequences of selection on *future* developments.

it earlier and even provided a solution in the essay of 1844. In the 1850s, he did develop several new ideas that led to an explicit and final formulation of his principle of divergence. He came to appreciate the dynamism of the living environment as the selecting force operative in speciation. That appreciation allowed him to maintain that natural selection was constantly working to shape individual differences into varieties and varieties into species. Darwin also believed those environmental forces could perform the same function of segregating groups from each other so that incipient varieties or species would not be swamped out by individuals bearing mediocre or unfavorable variations. He assumed that divergence, *as a kind of natural selection*, could overcome swamping effects insofar as it acted on extreme differences, simultaneously eliminating the intermediate or less fit varieties. Darwin seems to have been led in this direction by his own experience as a breeder of pigeons in the 1850s. To produce the morphologically distinctive varieties of pigeon, he, like other fanciers, would select from a large stock the individuals that expressed extreme traits. He presumed that nature operated in the same intelligent way as the pigeon fancier: nature selected from a large number of creatures just those individuals of quite divergent character, with the aim of producing distinctive races. Those favored races would thus gain the upper hand in securing a place in the economy of nature, just as the fancy races of pigeon secured a place in the breeder's coops.

Let me now answer explicitly the three questions I posed early in the chapter. Darwin thought of divergence as a kind of natural selection. The advantage it promised would be a more successful hold on a place in nature, with the derivative effect of more life in an extended area. And the new idea he brought to bear in the 1850s, the insight that struck him during his carriage ride, was that nature selected extremes.

The notion that nature might select extremes could only be sustained by the model of natural selection that Darwin assumed in his very early theorizing, certainly in the essays of 1842 and 1844, and that he retained in the *Big Species Book* and the *Origin of Species*: the model of an omniscient, intelligent selector that worked for the good of each creature, and that ultimately produced the ramifying features of the tree of life, much as the human breeder filled out the tree of pigeon varieties.

One might suppose that Darwin insinuated this model of an intelligent designer into his theory in order to ward off any negative reactions by religious critics. But the 1842 essay was not intended for a public viewing; at that point Darwin was simply trying to work out for himself the parameters of his theory and to become conceptually clear about how his theory would construct nature. What, then, would justify his assumptions about natural selection? I believe it was his religious understanding of the disposition of nature. Darwin meant it when he wrote his friend Asa Gray, shortly after publication of the *Origin*, that he was "bewildered" by charges that his book was irreligious; he protested that he "did not intended to write atheistically." He told Gray that he thought events in nature came about by "designed laws," of which natural selection would have been one.[85] This view is confirmed by a line he inserted into the *Big Species Book*'s comparison of human breeders to natural selection: "See how differently Nature acts! By nature, I mean the laws ordained by God to govern the Universe."[86] Though Darwin's theory heralded the inauguration of modern biology, he was nonetheless an early nineteenth-century thinker. In his *Autobiography*, he confessed that when he wrote the *Origin of Species*, he was convinced of "a First Cause having an intelligent mind in some degree analogous to that of man."[87] Darwin allowed his tenuous faith to slip away in the mid-1860s; he suggested that the term best capturing his own religious views was that coined by Huxley: agnostic. But the point to be made is simply that when he worked out his theory from 1837 to 1859, he was a theist who believed that the laws of nature, including natural selection, were designed by the Creator. Hence, the kind of intelligence and moral concern with which Darwin endowed natural selection had its ultimate source in that higher power.

What about the many passages in the *Origin* that seem to deny the Creator a role in the evolution of species? The answer is straightforward: Darwin only objected to the direct, seriatim intervention of the Deity, the Lord creating each species individually. He wished to explain, as a good scientist, that all the events in nature occurred as the result of laws constantly operating, of which natural selection was one. But these laws, as he frequently affirmed, were secondary causes imposed by God.[88] The laws thus bore the imprint of an all-powerful intelligence and moral actor.

Aside from Darwin's explicit belief that the laws of nature were expressions of secondary causes having a Divine intelligence as the primary cause, another factor may also have played a role. I offer this as a speculative consideration. In a set of reading notes on John Macculloch's *Proofs and Illustrations of the Attributes of God*, which were probably jotted down a short while after he had

85. Darwin to Asa Gray (22 May 1860), in *Correspondence of Charles Darwin*, 8:224.

86. Darwin, *Big Species Book*, 224.

87. Darwin, *Autobiography*, 92–93.

88. So, for example, Darwin, *Origin of Species*, 488: "To my mind it accords better with what we know of the laws impressed on matter by the Creator, that the production and extinction of the past and present inhabitants of the world should have been due to secondary causes."

read Malthus in late September 1838, Darwin linked the action of natural selection with the characteristic behavior of our own mentality, our own reason. In the elliptical note, he considered the hinge of a bivalve and compared it to what human intelligence could produce: "An adaptation made by intellect this process is shortened, but yet analogous [to operations of selection in nature], no savage ever made a perfect hinge.—reason, & not death rejects the imperfect attempts."[89]

Thus Darwin may well have been considering that human reason—certainly his own included—worked in the same way as natural selection: both made many trials until some trait or idea gave a small advantage. The difference between the two processes is that reason rejects unfit ideas, whereas natural selection rejects unfit individuals. Hence, the similarity of processes may have encouraged Darwin to think of natural selection as an intelligent process. Well, this is a bit of speculation.

My construction, admittedly, is not the common or established view of Darwin's conception of nature and its operations. The received view of his accomplishment is expressed, for example, by Lewontin, Rose, and Kamin: "Natural selection theory and physiological reductionism were explosive and powerful enough statements of a research program to occasion the replacement of one ideology—of God—by another: a mechanical, materialist science."[90] Most recently the received view has obtained a stamp of approval from Elliott Sober, who attempts to place Darwin's assertion that the laws of nature were promulgated by God into the category of Darwin's philosophical, as opposed to his scientific, views. Darwin, according to Sober, practiced methodological naturalism as we understand it today,[91] but the plain language of Darwin's *Origin of Species*, which embodies his theory, speaks otherwise.

When Fodor charges that contemporary Darwinian theory smuggles into the conception of natural selection an assumption that nature has intentional capacity, could Darwin's original construction be the source of the contraband? I hardly think so—there's no evidence that Fodor ever picked up the *Origin of Species*. Moreover, in subsequent editions of the book, Darwin attempted to amend some of the assumptions that seemed to rely on intentional discriminations by nature. He became sensitive to the problem when his friend Alfred Russel Wallace complained that the term "natural selection" was too

89. Charles Darwin, "Abstract of John Macculloch 1837 *Proofs and Illustrations of the Attributes of God*" (MS 58v), in *Charles Darwin's Notebooks*, 638.

90. Richard Lewontin, Steven Rose, and Leon Kamin, *Not in Our Genes* (New York: Pantheon, 1984), 51.

91. Elliott Sober, "Darwin and Naturalism," in Sober, *Did Darwin Write the Origin Backwards?* esp. 128.

anthropomorphic. One critic, Wallace reported, had observed that Darwin "manifestly endows 'Nature' with the intelligent faculty of designing and planning."[92] Darwin, as Wallace supposed, did not mean to suggest that idea, at least not by the mid-1860s. Darwin quickly agreed with his friend that Herbert Spencer's phrase "survival of the fittest" might equally serve, and he inserted those terms into the fifth edition of the *Origin* (1869)—though he retained the locution "natural selection," not wishing to give up an expression that captured his intentions so well.[93] When Fleeming Jenkin, one of the *Origin's* reviewers, forcefully insisted on the difficulties of the swamping problem,[94] Darwin suddenly realized the depth of the dangers. Fumbling for a response, he suggested, also in the fifth edition, that the environment might, in a Lamarckian way, produce individual variations all in the same direction; hence, natural selection would have the deck stacked, as it were, against swamping.[95] These adjustments may have mitigated the difficulties, but certainly did not eliminate them. If Darwin's theory is contained in the language of his book, then that theory depends on the ascription of intentions to nature—even though Darwin's own attitudes and beliefs became more astringent in his later years.[96]

Darwin's theory, of course, continued to evolve at the hands of subsequent generations of neo-Darwinists. Their manipulations drained the nineteenth-century spirit from the theory, leaving a more obviously mechanical framework—thus the contemporary appropriateness of referring to "the mechanism of natural selection." Fodor, then, would have been right had his objections been leveled at the theory as expressed in the *Origin of Species*. He took aim, however, at the agile neo-Darwinian theory and missed the more inviting target completely.

92. Alfred Russel Wallace to Darwin (2 July 1866), in *Correspondence of Charles Darwin*, 14:227–29. See also *a* review of *Origin of Species* in *Quarterly Journal of Science* 3 (1866): 151–76, quotation at 153. The reviewer concluded that Darwin was partially right, though the power of selection had to be in the decisions of the Deity.

93. Darwin to A. R. Wallace (5 July 1866), in *Correspondence of Charles Darwin*, 14:235–36.

94. Jenkin pointed out that Darwin's extremes were comparable to rare "sports," that is, large, favorable variations. In a normal population, however, a sport would naturally mate with those lacking the extreme trait, and with each generation the advantage would be diminished until it virtually vanished. See Fleeming Jenkin, "The Origin of Species," *North British Review* 46 (1867): 277–318, esp. 288–92.

95. Charles Darwin, *The Origin of Species: A Variorum Text*, ed. Morris Peckham (Philadelphia: University of Pennsylvania Press, 1959), 179. The 5th edition of the *Origin* countered Jenkin's review with "The conditions [of the environment] might indeed act in so energetic and definite a manner as to lead to the same modifications in all the individuals of the species without the aid of selection."

96. I've discussed Darwin's attribution of intelligence to nature from a different perspective in chapter 2 of this volume.

CHAPTER FOUR

# Darwin's Romantic Quest

## *Mind, Morals, and Emotions*

*If all men were dead, monkeys make men.—Men make angels.*

—Charles Darwin, *Notebook B*

From the beginning of his theorizing about species, Darwin had human beings in view. In the initial pages of his first transmutation notebook, he observed that "even mind & instinct become influenced" as the result of adaptation to new circumstances.[1] Considering matters as a Lyellian geologist, he supposed that such adaptations would require many generations of young, pliable minds being exposed to a changing environment. Captain FitzRoy had attempted to "civilize" the Fuegian Jemmy Button by bringing him to London and instructing him in the Christian religion; back in South America, however, Button reverted to his old habits, demonstrating to Darwin that the "child of savage not civilized man"—transmutation of mind was not the work of a day.[2] Darwin, though, had quickly become convinced that over long periods of time human mind, morals, and emotions had progressively developed out of savage origins. As he bluntly expressed it in his first transmutation notebook: "If all men were dead, monkeys make men.—Men make angels."[3] Presumably the transmutation of human beings into those higher creatures remained far in the future.

From July 1837, when he jotted these remarks in the first few pages of his *Notebook B*, to the early 1870s, with the publication of his *Descent of Man* and

1. Charles Darwin, *Notebook B* (MS p. 3), in *Charles Darwin's Notebooks, 1836–1844*, ed. Paul Barrett et al. (Ithaca: Cornell University Press, 1987), 171.

2. Ibid. (MS p. 4), 171.

3. Ibid. (MS pp. 169, 215), 213, 224. The second reference is a commentary on the first.

*Expression of the Emotions in Man and Animals*, Darwin gradually worked out theories of the evolution of human mentality that, in the main, we still accept. In the case of moral behavior, he produced a theory of its evolution that stands as a most plausible empirical account and displays the range and subtly of his genius. Examination of this history reveals that his conception of human mind had roots traversing a large swath of native ground. Some of those roots, though, extended to quite foreign soil, namely, German Romanticism.

## ON THE *BEAGLE* WITH HUMBOLDT

Darwin's conception of nature as well as his estimate of that smaller nature found in the human animal took definite shape during his five-year voyage on the *Beagle*. His experiences during the journey occurred within a framework already prepared by his enthusiastic reading of Alexander von Humboldt's *Personal Narrative of Travels to the Equinoctial Regions of the New Continent, 1799–1804*, a multivolume work that originally sparked his desire to sail to exotic lands.[4] Indeed, while a student at Cambridge he took to copying out long passages from the *Personal Narrative* and reading them to his patient friends. When he got the opportunity to embark on the *Beagle*, he brought along some of Humboldt's volumes as his vade mecum. Humboldt, a protégé of Goethe and friend of Schelling, represented nature in the Americas not as a stuttering, passionless machine that ground out products in a rough-hewn manner but as a cosmos of interacting organisms, a complex whose heart beat with law-like regularity, while yet expressing aesthetic and moral values. When Darwin first entered a South American jungle, he experienced its dazzling beauty and unique features in a Humboldtian mode, as he related in the diary he kept during his journey:

> I believe from what I have seen Humboldts glorious descriptions are & will forever be unparalleled: but even he with his dark blue skies & the rare union of poetry with science which he so strongly displays when writing on tropical scenery, with all this falls short of the truth. The delight on experiences in such times bewilders the mind. . . . The mind is a chaos of delight on experiences in a world of future & more quiet

4. Alexander von Humboldt and Aimé Bonpland, *Personal Narrative of Travels to the Equinoctial Regions of the New Continent, during the Years 1799–1804*, 7 vols., trans. Helen Williams (London: Longman, Hurst, Rees, Orme, and Brown, 1818–29).

pleasure will arise.— I am at present fit only to read Humboldt; he like another Sun illumines everything I behold.[5]

Humboldt's conception of nature resonated with the ideas of Goethe, whose views on metamorphosis he adopted, as well as with those of the Romantic philosopher Friedrich Schelling, who conceived nature and self to be linked at the deepest levels, a connection that left no room for that independent, personal God who displayed the inclinations of a Manchester industrialist. Darwin, of course, did not plunge below the surface of Humboldt's thought, but he nonetheless felt the power of the German's representation, which this diary entry, made during the voyage back to England, indicates: "As the force of impression frequently depends on preconceived ideas, I may add that all mine were taken from the vivid descriptions in the Personal Narrative which far exceed in merit anything I have ever read on the subject."[6]

Humboldt's name litters Darwin's diary and the book he made out of it, his *Journey of the Voyage of the Beagle* (1839). Humboldt's conception of nature as a creative force and a repository of moral and aesthetic values would lie at the foundation of all Darwin's later work on species, and especially the human species.[7] The creative force of nature would often, in Darwin's estimate, work through that most mundane yet transcendent faculty—instinct.

## EARLY THEORIES OF INSTINCT, EMOTION, AND REASON

The phenomenon of animal instinct would serve Darwin as the ground for understanding its outgrowth in human reason and moral behavior. He initially employed the conception of instinct, however, more generally in his explanation of species change. Prior to having read Malthus, he had formulated several theories to account for heritable modifications, the most prominent of which depended on the notions of use-inheritance and its result, instinct. Darwin assumed that in a changed environment, an animal might adopt habits that would accommodate it to the new conditions. Over many generations, these habits would, he believed, become instinctive, that is, expressed as innately determined behaviors. Such instincts, in time, would slowly alter anatomy, produc-

5. Charles Darwin, *Charles Darwin's Beagle Diary*, ed. R. D. Keynes (Cambridge: Cambridge University Press, 1988), 42.

6. Ibid., 443.

7. I have discussed in more detail Humboldt's contribution to Darwin's conception of nature in Richards, *The Romantic Conception of Life: Science and Philosophy in the Age of Goethe* (Chicago: University of Chicago Press, 2002), 518–26.

ing adaptive alterations, or so he supposed. This "view of particular instinct being memory transmitted without consciousness" had the advantage, he thought, of distinguishing his explanation of species change from Lamarck's, which he interpreted as appealing to a *conscious willing*—"Lamarck's willing absurd," he told himself.[8] Even after Darwin adopted natural selection as the principal means for producing species change, he still retained use-inheritance in his explanatory repertoire: it would become one of those sources for variation on which natural selection might work; in some instances, he would simply credit use-inheritance as the cause of an attribute that could not easily be explained by natural selection.

After he returned from his voyage, Darwin often visited the Zoological Society, where he had deposited for analysis and classification many of the animal specimens he had brought back on the *Beagle*; he thus had frequent occasion to visit the Society's menageries. During April 1838, he spent some time watching the apes and monkeys at the gardens, and he reflected on their emotional outbursts, which seemed to him quite humanlike. He was especially interested in an orangutan that "kicked & cried, precisely like a naughty child" when teased by its keeper.[9] In his notebooks, he placed such typical reactions within the framework of his theory of instinct: "Expression, is an hereditary habitual movement consequent on some action, which the progenitor did, when excited or disturbed by the same cause, which «now» excites the expression."[10] So, for example, Darwin speculated that the emotional response of surprise—raised eyebrows, retracted eyelids, and so forth—had arisen by association with our ancestors' efforts to see objects in dim light; now when the analogously unexpected object or event confronted us, we would react in an instinctual way, even though the light was perfectly adequate.[11] In this construction, the expression of emotion thus had no particular usefulness; it was understood, rather, as a kind of accidental holdover from the customary behavior of ancestors. Darwin would retain this basic notion about emotional display for the account he would later develop in the *Expression of the Emotions in Man and Animals* (1872). Emotional expression had its roots in instinct, and, in Darwin's view, reason did as well.

8. See Charles Darwin, *Notebook C* (MS p. 63), in *Charles Darwin's Notebooks*, 259.

9. Charles Darwin to Susan Darwin (1 April 1838), in *Correspondence of Charles Darwin*, ed. Frederick Burkhardt et al., 19 vols. to date (Cambridge: Cambridge University Press, 1985–), 2:80.

10. Charles Darwin, *Notebook M* (MS p. 107), in *Charles Darwin's Notebooks*, 545. Double wedges indicate a later insertion by Darwin.

11. Ibid. (MS p. 95), 542.

In August 1838, Darwin began reading David Hume's *Inquiry concerning Human Understanding*.[12] Hume's representation of ideas as less vivid copies of sensations perfectly accorded with Darwin's intuitions about the continuity of animal and human mentality, for if ideas were copies of impressions, animals would be quite capable of thought. Darwin developed this sensationalist epistemology in his *Notebook N*, where he proposed that simple reasoning consisted in the comparison of sensory images and that the recollection of several such images producing a pleasant state was of the very nature of complex thought: "Reason in simplest form probably is single comparison by sense of any two objects—they by VIVID power of conception between one or two absent things.—reason probably mere consequence of vividness & multiplicity of things remembered & the associated pleasure as accompanying such memory."[13]

Just as Hume understood reason to be a kind of "wonderful and unintelligible instinct in our souls," so Darwin as well thought of intellectual activity to be a "modification of instinct—an unfolding & generalizing of the means by which an instinct is transmitted."[14] Human intelligence was thus not opposed to animal instinct but grew out of it in the course of ages.

In finding the antecedents of human rationality in animal sources, Darwin really opened no new epistemological ground. Carl Gustav Carus, Goethe's disciple and an author whom Darwin read in early 1838, asserted the decidedly romantic thesis that mind and matter ran together throughout nature. Adopting Carus's language, Darwin contemplated a nature alive with mind. He reflected that "there is one living spirit, prevalent over this world . . . which assumes a multitude of forms according to subordinate laws." And like Carus, he concluded that "there is one thinking . . . principle intimately allied to one kind of matter—brain" and that this thinking principle "is modified into endless forms, bearing a close relation in degree and kind to the endless forms of the living beings."[15] Darwin's assumption of cognitive continuity between men and animals would not even have offended the religiously minded among

12. See Darwin's remarks, ibid. (MS p. 104), 545; and Charles Darwin, *Darwin's Reading Notebooks*, in *Correspondence of Charles Darwin*, 4:438.

13. Charles Darwin, *Notebook N* (MS p. 21e), in *Charles Darwin's Notebooks*, 569.

14. David Hume, *Treatise of Human Nature*, ed. L. A. Selby-Bigge (Oxford: Clarendon Press, [1739] 1888), 179. Darwin refers to this passage in his *Notebook N* (MS p. 101), 591, and remarks, "Hume has section (IX) on Reason of Animals . . . he seems to allow it is an instinct." Darwin, *Notebook N* (MS p. 48), 576.

15. Darwin, *Notebook C* (MS p. 210e), 305. I read "world" for the transcription "word." Darwin studied Carus in translation. See Carl Gustav Carus, "The Kingdoms of Nature, Their Life and Affinity," *Scientific Memoirs* 1 (1837): 223–54.

his own countrymen. Several natural theologians whom he read during the late 1830s and early 1840s—John Fleming, Algernon Wells, and Henry Lord Brougham, for instance—did not blanch to find some glimmer of reason exhibited even among the lower animals.[16] But no animal, in the estimation of these British writers, gave evidence of any hint of what was truly distinctive of human mind—namely, moral judgment. If Darwin were to solidify his case for the descent of man from lower animals, he would have to discover the roots of moral behavior even among those creatures. And so he did.

## MORAL THEORY PRIOR TO THE *ORIGIN OF SPECIES*

Darwin's own moral sensitivities were assaulted during his South American travels, especially by the Brazilian slave trade. His family cultivated strong abolitionist sentiments, which originated with both of his grandfathers; his sisters kept him informed about the efforts in Parliament to emancipate the slaves in the British colonies.[17] Darwin had his convictions reinforced by the many observations Humboldt himself had made about the loathsome trade in human beings.[18]

Darwin's own fury could be barely suppressed when he witnessed African families being separated at slave auctions and slaves being beaten and

16. See especially John Fleming, *The Philosophy of Zoology*, 2 vols. (Edinburgh: Constable, 1822), 1:220–22; Algernon Wells, *On Animal Instinct* (Colchester: Longman, Rees, Orme, Brown, Green, and Longman, 1834), 20; and Henry Lord Brougham, *Dissertations on Subjects of Science concerned with Natural Theology: Being the Concluding Volumes of the New Edition of Paley's Work* (London: Knight, 1839), 175. Darwin's copy of Fleming, with annotations, is held in the Department of Manuscripts, Cambridge University Library. His notes on Brougham and Wells are in his *Notebook N* (MS pp. 62, 62–72), 580, 582–84, respectively. He wrote: "Lr. Brougham . . . says animals have abstraction because they understand signs.—very profound.—concludes that difference of intellect between animals & men only in Kind [*sic*; Degree]."

17. Darwin's sister Susan kept him abreast of the Parliamentary debates. See Susan Darwin to Charles Darwin (3–6 March 1833), in *Correspondence of Charles Darwin*, 1:299. For a discussion of Darwin's antislavery attitude and the range of sentiments about slavery held by intellectuals of the period, see Adrian Desmond and James Moore, *Darwin's Sacred Cause* (New York: Houghton Mifflin Harcourt, 2009). Desmond and Moore admirably show the range of pro- and antislavery attitudes entertained by Darwin's contemporaries. They ascribe Darwin's conviction of the unity of humankind to his antislavery attitudes, and they argue that this assumption of unity led to his theory of species descent. There is, however, no reason to think that a belief in the unity of mankind would be a defense against slavery—Christian slaveholders of the American South quite easily entertained the notion of human unity while subjugating one branch of their own species. Moreover, Darwin's notebooks suggest that he first conceived of descent in terms of animals, not humans. See Robert J. Richards, "The Descent of Man: Review of *Darwin's Sacred Cause*," *American Scientist* 97 (September–October 2009): 415–17.

18. See Humboldt and Bonpland, *Personal Narrative*, 3:3.

degraded. He recalled poignantly in his *Journal of Researches of the Voyage of the Beagle* an incident that powerfully illustrated for him the character of the peculiar institution. He was on a ferry with an African slave, who did not understand English. Darwin, typical perhaps of the Englishman abroad, gesticulated and raised his voice to make himself understood.

> He, I suppose, thought I was in a passion, and was going to strike him; for instantly, with a frightened look and half-shut eyes, he dropped his hands. I shall never forget my feelings of surprise, disgust, and shame, at seeing a great powerful man afraid even to ward off a blow, directed, as he thought, at his face. This man had been trained to a degradation lower than the slavery of the most helpless animal.[19]

When finally the *Beagle* left Brazil, Darwin rejoiced that "I shall never again visit a slave-country." He perceived immediately that utilitarian motives would do little to suppress this kind of evil:

> It is argued that self-interest will prevent excessive cruelty; as if self-interest protected our domestic animals, which are far less likely than degraded slaves, to stir up the rage of their savage masters. It is an argument long since protested against with noble feeling, and strikingly exemplified, by the ever illustrious Humboldt.[20]

This last remark about the deficiencies of utilitarian considerations to adjudicate moral responsibility came in the revised edition (1844) of Darwin's *Journal of Researches*. Prior to this time, he had made an effort to found an initial hypothesis about the evolution of morals on utilitarian grounds.

Darwin knew quite well William Paley's *Moral and Political Philosophy* (1785) from his undergraduate days at Cambridge. Now, while exploring the various branches of his developing theory in early September 1838, he momentarily adopted Paley's central rule of "expediency." This rule grounded moral approbation in what, in the long run, would be useful, that is, beneficial either to an individual or a group and, as a consequence, would supply the pleasure

19. Charles Darwin, *Journal of Researches into the Geology and Natural History of the Various Countries Visited by H.M.S.* Beagle *under the Command of Captain FitzRoy, R.N., from 1832 to 1836* (London: Henry Colburn, 1839), 28.

20. Charles Darwin, *The Voyage of the Beagle*, ed. Leonard Engel, 2nd ed. (New York: Doubleday, [1844] 1962), 497.

God intended for mankind.[21] Darwin gave this rule a biological interpretation: "Sept 8th. I am tempted to say that those actions which have been found necessary for long generation, (as friendship to fellow animals in social animals) are those which are good & consequently give pleasure, & not as Paley's rule is then that on long run *will* do good.—alter *will* in all such cases to *have & origin* as well as rule will be given."[22] Darwin here suggested that those habits that preserved animals—such as friendship and nurture of young—must have been practiced over many generations and so became instinctive. What we call "good," then, are those long-term, beneficial instincts that have proved necessary for social cohesion and development. Hence, Darwin supposed that what Paley took to be a forward looking rule—act to achieve general utility in the *future*—might be transformed into one describing instincts that arose from social behaviors that had been beneficial over long periods in the *past*. But this biologized Palyean ethics receded from Darwin's purview after he examined a volume containing a more penetrating analysis of morals, James Mackintosh's *Dissertation on the Progress of Ethical Philosophy* (1836).

The young Darwin knew Mackintosh personally—he was the brother-in-law of Darwin's uncle Josiah Wedgewood—and thought of this distant relation as "the best converser on grave subjects to whom I have ever listened."[23] In his *Dissertation*, Mackintosh objected to Paley's notion that selfish pleasure ultimately motivated right action. He rather sided with the likes of Shaftesbury, Butler, and Hutchinson, who believed that human nature came outfitted with a deep sense of moral propriety. Human beings, Mackintosh maintained, acted spontaneously for the welfare of their fellows and immediately approved of such actions when displayed by others. Yet he did not deny the utility of moral conduct. In a cool hour we could assess moral behavior and rationally calculate its advantages, but such calculation was not, he thought, the immediate spring of action, which lay coiled in the human soul. Mackintosh thus distinguished the *criterion* for right conduct—utility—from the *motive* for such conduct—an innate disposition.

This analysis fit smoothly into Darwin's developing conception of moral behavior, a conception that both appreciated the utility of ethical behavior and recognized its deep biological roots. Darwin's notes on Mackintosh's

21. William Paley, *The Principles of Moral and Political Philosophy*, 2 vols., 16th ed. (London: R. Faulder, 1806), 1:89–90, 76.

22. Darwin, *Notebook M* (MS p. 132e), 552. Darwin made a similar observation about Paley's rule of utility in Darwin, *Old and Useless Notes* (MS pp. 50–51), in *Charles Darwin's Notebook*, 623.

23. Charles Darwin, *The Autobiography of Charles Darwin*, ed. Nora Barlow (New York: Norton, 1969), 66.

*Dissertation* reveal, however, that he discovered a jarring patch in the original theory, but one which he believed his own biological approach could pave over. The difficulty was this: What explained the harmony of the criterion for moral conduct and the motive for such behavior? Why were we moved to act spontaneously in a way that we might later, in a moment of reflection, recognize to have social utility? Not impressed with Mackintosh's faint appeal to a divine harmonizer, Darwin suggested that the innate moral knowledge we harbored was really an instinct acquired by our ancestors. The instinct did indeed have social utility, but like all instincts it also had a motivational urgency not connected with any rational calculation of pleasures and pains. The instinct thus bound up both the criterion and the motive. Such instincts, Darwin thought, would be sufficiently different from our other more abrupt and momentary instincts in that they would be persistent and firm and thus evoke a more reverential feeling.

Darwin moved with alacrity along this line of thought because in this instance, as in many others, he found that his theory of biological development solved a problem that remained loose and frayed in the humanistic literature. On 3 October 1838, a few days after Malthus furnished the key stimulus to the idea of natural selection, the young biologist reformulated his theory of moral conscience along the lines suggested by Mackintosh.[24] Darwin assumed that habits of parental nurture, group cooperation, community defense, and the like would be sustained over many generations, driving such habits into the heritable legacy of a species, so that they would be manifested in succeeding generations as instincts for moral conduct. These instincts would be distinguished from fleeting inclinations and less persistent impulses, which might occur in one generation and depart with the next. When an individual with sufficient intelligence recalled, in a cool hour, a behavior elicited by these deeply ingrained dispositions, he or she would feel renewed satisfaction and also would be able to perceive on reflection the social utility of the behavior. Darwin thus solved the problem of the coincidence of the moral motive and the moral criterion.

Darwin worked out the basic framework of his moral conception without aid of the theory he had recently formulated, namely, natural selection. When he began to apply the device of natural selection to explain instincts, however, he stumbled at the brink of a yawning conceptual abyss, which threatened to swallow his entire theory of evolution by natural selection. The crucial diffi-

24. Darwin, *Notebook N* (MS pp. 1–3), 536.

culty was this: the social instincts most frequently gave advantage to the recipients of moral actions, not to their agents; natural selection, however, preserved individuals because of traits advantageous to themselves, not to others. Darwin first met this difficulty when studying the social insects in the 1840s, when the problem became even more complicated.

Soldier bees and ants displayed anatomical traits and instinctive behaviors that served the welfare of their colonies, not directly themselves. Indeed, a soldier bee might defend the hive at the cost of its own life. Moreover, these insects were neuters; consequently, they could not in the first instance pass beneficial adaptations to succeeding generations. How then could their other-regarding traits be explained, and, more generally, how did the attributes of neuters arise? Darwin worried about this problem for some time, fearing it would allow the Creator a return to those provinces from which he had been lately banished.[25] Only during the first months of 1858, while laboring on the manuscript that would become, in its abridged form, the *Origin of Species*, did Darwin discover the solution to his problem:

> I have stated that the fact of a neuter insect often having a widely different structure & instinct from both parents, & yet never breeding & so never transmitting its slowly acquired modifications to its offspring, seemed at first to me an actually fatal objection to my whole theory. But after considering what can be done by artificial selection, I concluded that natural selection might act on the parents & continually preserve those which produced more & more aberrant offspring, having any structures or instincts advantageous to the community.[26]

Thus the soldier bee that sacrificed its life for the hive would have had its instincts honed over generations not by individual selection but by natural selection, preserving those hives that had individuals with traits that benefited the entire community. With this account, which he reiterated in the *Origin of Species*, Darwin had the key to the puzzle of human moral action: as he would argue in the *Descent of Man*, altruistic impulses of individual members would

25. I have discussed this problem as well as other aspects of the development of Darwin's moral theory more extensively in Richards, *Darwin and the Emergence of Evolutionary Theories of Mind and Behavior* (Chicago: University of Chicago Press, 1987), chaps. 2 and 5.

26. Charles Darwin, *Charles Darwin's Natural Selection: Being the Second Part of His Big Species Book Written from 1856 to 1858*, ed. R. C. Stauffer (Cambridge: Cambridge University Press, 1975), 510.

give a tribal clan advantages over other clans, and thus such instincts would become characteristic of evolving human communities.

## THE PROBLEM OF HUMAN EVOLUTION, 1859–1871

In the late 1860s, Darwin initially approached the problem of human evolution quite modestly. He had originally intended to consider human beings only from the point of view of sexual selection, which he thought could explain the different attributes of males and females of the many races of mankind. He engorged the second part of *The Descent of Man and Selection in Relation to Sex* (1871) with detailed discussions of sexual selection throughout the animal kingdom, with only the last two substantive chapters devoted to human sexual dimorphism and racial differences. He argued that male combat for females among our ancestors would have contributed to the male's larger size, pugnacity, strength, and intelligence. The particular features of female beauty in the different races—generally hairless bodies, cast of skin, shape of nose, form of buttocks, and so forth—he thought would have arisen from male choice. Women generally displayed the tender virtues, but their intellectual attainments would be largely due, Darwin thought, to inheritance from the male parent. In a letter to a young American female college student, he ventured that if women went to university and were schooled over generations as the sons of the gentry were, then they would, through use-inheritance, become as intelligent as men. But were this to happen, "we may suspect that the easy education of our children, not to mention the happiness of our homes, would in this case greatly suffer."[27] Darwin's cultural attitudes did not stray far from those of the mid-Victorian gentleman.

Several events occurred during the 1860s that caused Darwin to alter the limited intentions he had for his book. Early in the decade, his great friend Charles Lyell waded into the undulating opinions forming about human evolution in the wake of the *Origin*. But the hedging argument of Lyell's *Antiquity of Man* (1863), which displayed a style cultivated at the Old Bailey, drove Darwin to distraction. Although Lyell admitted the physical similarity of human beings to other primates, he yet argued that the mental and moral constitution of humans placed them far above any other animals in the scale of being. Linguistic ability in particular demonstrated the wide gulf separating the mind of

27. Charles Darwin to Caroline Kennard (9 January 1882), DAR 185, Department of Manuscripts, Cambridge University Library.

man from that of animals. This was no chasm that could be bridged in "the usual course of nature." The move from animals to man, Lyell intimated, had to be carried on the wings of a divine spirit.[28]

Alfred Russel Wallace initially stood ready to combat Lyell's theological construction of human mind and morals. In a lecture delivered to the Anthropological Society in 1864, he produced an ingenious defense of the naturalistic position. He argued that natural selection, operating on our animal forbearers, produced the various races of men, though not yet their distinctive mental and moral characters. Only after these races appeared would natural selection operate on the various clans and tribes, preserving those groups in which individuals displayed sympathy, cooperation, and "the sense of right which checks depredation upon our fellows."[29]

Three features of Wallace's account of the evolution of human mind and morals stand out. First, he conceived the selective environment to be other protohuman groups—which would have an accelerating effect on the evolutionary process, since social environments would rapidly change through responsive competition. Second, he proposed that selection worked on the group, rather than the individual—which allowed him to explain the rise of altruistic behavior, that is, behavior perhaps harmful to the individual but beneficial to the group. In his original essay on the transmutation of species (1858), Wallace conceived of the struggle for existence to occur among *varieties* instead of individuals.[30] He continued to think in such group terms when considering the evolution of moral behavior. Finally, in a note to the published version of his talk to the Anthropological Society, he mentioned that he was inspired to develop his thesis by reading Herbert Spencer's *Social Statics*.[31] Spencer's own early brand of socialism had pulled Wallace to his side. In *Social Statics* (1851), Spencer had envisioned a gradual and continual adjustment of human beings to the requirements of civil society, with individuals accommodating themselves to the needs of their fellows, so that eventually a classless society would emerge in which the greatest happiness for the greatest number would

28. Charles Lyell, *The Geological Evidences of the Antiquity of Man* (London: Murray, 1863), 505.

29. Alfred Russel Wallace, "The Origin of Human Races and the Antiquity of Man Deduced from the Theory of 'Natural Selection,'" *Anthropological Review* 2 (1864): clxiii.

30. See Charles Darwin and Alfred Russel Wallace, "On the Tendency of Species to Form Varieties, and on the Perpetuation of Varieties and Species by Natural Means of Selection," read 1 July 1858, *Journal of the Proceedings of the Linnaean Society of London, Zoology* 3 (1858):45–62.

31. Wallace, "Origin of Human Races and the Antiquity of Man," clxx.

be realized.[32] Spencer assumed that the inheritance of useful habits would be the means by which such evolutionary progress would occur, while Wallace believed natural selection to be the agent of that progress.

Darwin welcomed Wallace's solution to the evolution of human morality, since he himself had developed certain views about community selection in social insects congenial to his friend's position. Darwin would emphasize, however, that the members of small tribes, of the sort Wallace envisioned, would likely be related; therefore, a disadvantage to a given individual practicing altruism would yet be outweighed by the advantage of the practice to recipient relatives. Ultimately, however, Darwin dropped this qualification and simply embraced group selection as operative in human (and animal) societies.[33]

Wallace's faith in a naturalistic account of human evolutionary progress, however, succumbed to the evidence of higher powers at work in the land. Though raised as a materialist and agnostic, Wallace had chanced to attend a séance, which piqued his empiricist inclinations. Shortly thereafter, in 1866, he hired a medium in order to investigate the phenomena usually attendant on the invocation of the spirit world. Wallace, gentle soul that he was, became a true believer (unlike Darwin, who regarded spiritualism as rubbish). Wallace's new conviction focused his attention on certain human traits—naked skin, language, mathematical ability, ideas of justice, and abstract reasoning generally—that would confer no biological advantage on individuals in a low state of civilization. Indeed, Wallace believed that for sheer survival, human beings needed a brain no larger than that of an orangutan, or perhaps one comparable to that of the average member of a London gentleman's club. Such traits as abstract reasoning and moral sensitivity, therefore, could not be explained by natural selection. Yet in both aboriginal and advanced societies, individu-

32. Herbert Spencer, *Social Statics: or, The Conditions Essential to Human Happiness Specified and the First of Them Developed* (London: Chapman, 1851). Spencer's own trajectory moved from an early, youthful enthusiasm for communism to the laissez-faire individualism of his later years. See Richards, *Darwin and the Emergence of Evolutionary Theories of Mind and Behavior*, chaps. 6 and 7, for a discussion of Spencer's development.

33. Darwin generalized his concept of community selection to embrace what we would call group selection—that is, selection of groups of individuals for traits that benefit the group, even if its members are not related. One can trace Darwin's thinking about this in a passage from the *Origin* that changed considerably over the book's six editions. In the first edition, the passage reads: "In social animals it [natural selection] will adapt the structure of each individual for the benefit of the community; if each in consequence profits by the selected change"; Darwin, *On the Origin of Species* (London: Murray, 1859), 87. The fifth edition (1869) altered the last phrase to say: "if this in consequence profits by the selected change." And the sixth edition (1872) more clearly puts it: "if the community profits by the selected change." For the several passages, see Charles Darwin, *The Origin of Species: A Variorum Text*, ed. Morse Peckham (Philadelphia: University of Pennsylvania Press, 1959), 172. It appears that Darwin's work on human group selection in the *Descent of Man* led him to generalize his concept in the last editions of the *Origin*.

als displayed these qualities. While his friend Herbert Spencer regarded such properties as explicable only through use-inheritance, Wallace found a unique explanatory mode of selection that his new faith could provide.[34] In his estimation, distinctively human traits had been artificially selected for us: "a superior intelligence," he proposed, "has guided the development of man in a definite direction, and for a special purpose, just as man guides the development of many animal and vegetable forms."[35] We were thus like domestic creatures in the hands of higher spiritual powers, and they artificially selected distinctively human traits for our advantage.

When Darwin learned of Wallace's turnabout, he was dumbfounded: "But I groan over Man—you write like a metamorphosed (in the retrograde direction) naturalist, and you the author of the best paper that ever appeared in the *Anthropological Review*! Eheu! Eheu! Eheu!"[36] Though Wallace's flight to other powers than nature was fueled by his new faith, the crux of his argument had force: Since natural selection operated only on traits that provided some immediate biological advantage, how might one explain human traits that seemed not particularly useful, at least for survival.

Another writer, friendly to the Darwinian cause, yet spied a comparable problem in the assumption of human evolutionary progress. William Rathbone Greg, Scots moralist and political writer, discovered that a keen moral sense might spread seeds of wicked growth. A highly civilized society, he remarked, would be inclined to protect not only the physically weak from the winnowing hand of natural selection but the intellectually and morally degenerate as well. So protected, the inferior types would have opportunity to out breed their betters. Greg, a Scots gentleman of refined sensibility, regarded the case of the Irish as cautionary:

> The careless, squalid, unaspiring Irishman, fed on potatoes, living in a pig-sty, doting on a superstition, multiplies like rabbits or ephemera:—the frugal, foreseeing, self-respecting, ambitious Scot, stern in his

34. Spencer contended that the higher mental powers required delicate coadaptation of elemental traits that themselves could have provided no advantage singly. Moreover, many mental powers—aesthetic preference, for instance—had no survival value at all and could not therefore have arisen by natural selection. See Herbert Spencer, *Principles of Biology*, 2 vols. (New York: D. Appleton, [1867] 1884), 454–55. Wallace wrote Darwin (18 April 1869) to say that his altered view about human evolution derived from his empirical testing of the medium's power. See James Marchant, *Alfred Russel Wallace: Letters and Reminiscences*, 2 vols. (London: Cassell, 1916), 1:244.

35. Alfred Russel Wallace, "The Limits of Natural Selection as Applied to Man" (1870), in Russel, *Natural Selection and Tropical Nature* (London: Macmillan, 1891), 204.

36. Darwin to Alfred Wallace (26 January 1870), in *Correspondence of Charles Darwin*, 18:17.

> morality, spiritual in his faith, sagacious and disciplined in his intelligence, passes his best years in struggle and in celibacy, marries late, and leaves few behind him. . . . In the eternal "struggle for existence," it would be the inferior and less favoured race that had prevailed—and prevailed by virtue not of its good qualities but of its faults.[37]

The profligate and degenerate Irish yet seemed to be winning the evolutionary race by the trait that counted—reproduction. The considerations of Lyell, Wallace, and Greg spurred Darwin to expand his intended volume on sexual selection to tackle these apparent barriers to a naturalistic understanding of human evolution.

## MIND AND MORALS IN THE *DESCENT OF MAN*

In the face of Greg's argument, Darwin collected in the *Descent of Man* considerable evidence about the fortunes of the reprobate Irish. On the basis of this evidence, he maintained that many natural checks to the less fit would ultimately forestall their advance: the debauched would suffer higher mortality, criminals would sire fewer offspring, and the wicked would likely die young. Yet it could be that the likes of the Irish, though decidedly less able, would simply crowd out the British. After all, though evolutionary progress was general, it was "no invariable rule."[38]

In his response to Greg's concern, Darwin made an implicit distinction between the *meaning* of fitness—that is, certain properties, such as high intelligence and moral judgment—and the *criterion* of fitness—that is, survival and reproductive success. Were meaning and criterion collapsed into one, then the principle of natural selection would have devolved into a tautology: the fit survive, and by the fit we mean the survivors. Darwin's original conception of natural selection asserted that fitness traits had causal consequences, that is, survival. But his conception certainly allowed that those causal consequences might, for contingent reasons, fail. Progress is no invariable rule.

Lyell's and Wallace's objections to the application of natural selection in the case of man proved more difficult than that of Greg, but they brought Darwin to several ingenious solutions to the problems posed. Linguistic ability stood chief among the features of intelligence that had to be considered. In dealing

37. [William R. Greg], "On the Failure of 'Natural Selection' in the Case of Man," *Fraser's Magazine* 78 (1868): 353–62, quotation at 361. Darwin quotes this passage with some relish in the *Descent of Man and Selection in Relation to Sex*, 2 vols. (London: Murray, 1871), 1:174.

38. Darwin, *Descent of Man*, 1:174–80, quotation at 177.

with this problem, Darwin reverted to a theory he had initially entertained in his *Notebook N*, which he had kept between 1838 and 1839. There he had sought to develop a naturalistic account of the origin of language. He supposed that our aboriginal ancestors began imitating sounds of nature (e.g., "crack," "roar," "crash") and that language developed from these simple beginnings.[39] In the late 1860s, while working on the *Descent*, Darwin made frequent inquiries of his cousin, the linguist Hensleigh Wedgwood, who likewise advanced the onomatopoetic theory of language. Darwin also relied on another book in formulating his thesis about the function of language in human evolution. This was by a German linguist, August Schleicher, a friend of Ernst Haeckel and a new convert to Darwinian theory. In his *Darwinsche Theorie und die Sprachwissenschaft* (Darwinian theory and the science of language, 1863), Schleicher maintained that contemporary languages had gone through a process in which simpler *Ursprachen* had given rise to descendant languages that obeyed natural laws of development of the kind Darwin had proposed for biological organisms.[40] He argued that Darwin's theory was thus perfectly applicable to languages and, indeed, that evolutionary theory itself was confirmed by the facts of language descent (see chap. 8 in this volume). With these linguistic resources, Darwin had a counterargument to Wallace's, one by which he could solidify an evolutionary naturalism.

Darwin conceded that Wallace had been correct: for sheer survival, our animal ancestors had sufficient brain power. But he could now blunt the further implication of his friend's argument. Citing Schleicher, he argued in the *Descent* that over the course of ages, the acquisition and development of language would rebound on brain, producing more complex trains of ideas; and constant exercise of intricate thought would gradually alter brain structures, causing a hereditary transformation and, consequently, a progressive enlargement of human intellect beyond that necessary for mere survival.[41]

39. See Darwin, *Notebook N* (MS p. 65), 581.

40. August Schleicher, *Die Darwinsche Theorie und die Sprachwissenschaft* (Weimar: Böhlau, 1863).

41. Darwin, *Descent of Man*, 1:57. In the conclusion to the *Descent of Man*, Darwin referred to an article by Chauncey Wright, which he had just read in the last moments of manuscript preparation. Wright had attacked Wallace's argument that man's big brain had to be given a nonselectionist account. See Chauncey Wright, "Limits of Natural Selection," *North American Review* 111 (October 1870): 282–311. Darwin suggested that Wright also endorsed the idea that language operated to produce man's increased intellectual capacity through use-inheritance (*Descent of Man*, 2:390–91). Wright's argument is a bit convoluted, but it is clear that in fact he did not make the argument Darwin attributed to him. Quite the contrary. Wright (294–98) maintained that Wallace had simply misjudged the character of the native's capacities. Wright rather held that language and so-called higher faculties were merely collateral features of capacities directly useful to the native and so indirectly acquired through natural selection. "Why may it not be," he asked,

Darwin's general theory of the rise of human intellect thus depended on the inheritance of acquired characteristics, or at least that is one of the strands of argument he employed. But it was not the only strand. Darwin's explanations in the *Origin* and the *Descent* were rhetorically robust—if the reader did not like one line of consideration, the author was ready with another line. His second strand of argument relied on community selection. In the *Descent*, Darwin contended that if a tribe of our aboriginal ancestors contained among its members some mute, inglorious Newton, an individual who through inventiveness and intellectual prowess benefited his tribe in competition with other tribes, then he and his relatives would survive and reproduce: "If such men left children to inherit their mental superiority, the chance of the birth of still more ingenious members would be somewhat better, and in a very small tribe decidedly better. Even if they left no children, the tribe would still include their blood-relations; and it has been ascertained by agriculturists that by preserving and breeding from the family of an animal, which when slaughtered was found to be valuable, the desired character has been obtained."[42] Darwin enunciated here an idea that in our time has become known as "inclusive fitness." A heritable trait that confers little or no benefit on an individual but sufficiently advances the cause of relatives will be preserved and spread as the group enlarges and forms daughter groups. Darwin first developed this theory of community selection to solve the problem of the evolution of the social insects; it now became the key to understanding the evolution of social human beings.

In the first volume of the *Descent*, the question of human moral judgment occupied the greatest measure of Darwin's attention. Moral sense was by common consent that attribute most distinctive of human beings. Both Lyell and Wallace could not conceive that a refined moral sense might have arisen naturally from animal stock. After all, moral behavior did not prove particularly beneficial to those exercising it—hence natural selection could not account for it. In explaining the rise of moral behavior, Darwin once again depended upon his theory of community selection and, one presumes, some hints derived from Wallace's lecture to the Anthropological Society. Darwin put it this way:

---

"that all that he [the savage] can do with his brains beyond his needs is only incidental to the powers which are directly serviceable?" (295). He further suggested that the difference between the savage and the philosopher "depends on the external inheritances of civilization, rather than on the organic inheritances of the civilized man" (296). Darwin, in his enthusiasm for the Schleicher argument, found its ghost in any text that opposed Wallace's thesis.

42. Darwin, *Descent of Man*, 1:161.

> It must not be forgotten that although a high standard of morality gives but a slight or no advantage to each individual man and his children over the other men of the same tribe, yet an increase in the number of well-endowed men will certainly give an immense advantage to one tribe over another. There can be no doubt that a tribe including many members who, from possessing in a high degree the spirit of patriotism, fidelity, obedience, courage, and sympathy, were always ready to give aid to each other and to sacrifice themselves for the common good, would be victorious over most other tribes; and this would be natural selection. At all times throughout the world tribes have supplanted other tribes; and as morality is one element in their success, the standard of morality and the number of well-endowed men will thus everywhere tend to rise and increase.[43]

Community selection proved an ingenious way to understand the evolution of human altruism. It yet had its own difficulty: How do these moral traits arise *within* one tribe in the first place? After all, as Darwin noted, it is not likely that parents of an altruistic temper would raise more children than those of a selfish attitude. Moreover, those who were inclined to self-sacrifice might leave no offspring at all.[44] Darwin employed his device of use-inheritance to explain the origin of such social behaviors within a given tribe. He proposed two related sources for such behaviors. The first is the prototype of our own, contemporary theories of reciprocal altruism. Darwin observed that as the reasoning powers of members of a tribe improved, each would come to learn from experience "that if he aided his fellow-men, he would commonly receive aid in return." From this "low motive," as he regarded it, each might develop the habit of performing benevolent actions, which habit might be inherited and thus furnish suitable material on which community selection might operate. The second source relied on the assumption that "praise and blame" of certain social behaviors would feed our animal need to enjoy the admiration of others and to avoid feelings of shame and reproach. This kind of social control would also lead to heritable habits.[45]

One salient objection to any theory of the biological evolution of moral conduct points to the often very different standards of acceptable behavior

43. Ibid., 166.
44. Ibid., 163.
45. Ibid., 163–65.

in various cultures. Darwin recognized that what might be approved as moral in one age and society might be execrated at a different time and place. The Fuegian Indians might steal from other tribes without the slightest remorse, whereas an English gentleman would regard such behavior with contempt. Nonetheless, members of these vastly different cultures would commonly endorse the obligation to deal sympathetically and benevolently with members of their own particular group. The English gentleman and lady, having more advanced intellects, would have learned that tribal and national differences were superficial, and thus they would have perceived a universal humanity underlying inessential traits. Their own instinctive sympathies would have thus been trained to respond to all human beings as members of a common tribe. In Darwin's conception, then, evolution would have molded the most primitive human beings to react altruistically to brothers and sisters; over the ages, however, cultural learning, coupled with increased intelligence, would reveal just who those brothers and sisters might be.[46]

"Philosophers of the derivative school of morals" (e.g., Bentham and Mill), Darwin observed, "formerly assumed that the foundations of morality lay in a form of Selfishness; but more recently in the 'Greatest Happiness principle.' "[47] Virtually all scientists and philosophers today who have considered the matter have located these utilitarian principles at the foundation of an evolutionary construction of ethics. Michael Ghiselin provides the prototypical example. He has argued that, according to Darwin's theory, since a so-called altruistic act furthers the competitive ability of self and family, that act is "really a form of ultimate self-interest."[48] Richard Dawkins, a defender of Darwin, yet warned "that if you wish, as I do, to build a society in which individuals cooperate generously and unselfishly towards a common good, you can expect little help from biological nature."[49] These sentiments, obviously, do not reflect Darwin's own view. Our moral instincts, he believed, would urge us to act for the benefit of others without calculating pleasures and pains for self. And since such altruistic impulses, at least in advanced societies, would not be confined to family, tribe, or nation, he confidently concluded that his theory removed "the reproach of laying the foundation of the most noble part of our nature in the base principle of selfishness."[50]

46. Ibid., 100–101.
47. Ibid., 97.
48. Michael Ghiselin, "Darwin and Evolutionary Psychology," *Science* 179 (1973): 967.
49. Richard Dawkins, *The Selfish Gene* (Oxford: Oxford University Press, 1976), 3.
50. Darwin, *Descent of Man*, 1:98.

## THE EXPRESSION OF THE EMOTIONS

Although Darwin believed our intelligence and moral responses had their roots in animal mind, he granted these faculties had yet developed far beyond those of our progenitors. By contrast, he considered human emotions and their display not to have comparably progressed. The fear displayed by his little dog over a wind-blown parasol differed little, he thought, from that of the native who trembled because invisible spirits might be causing a lightning storm—or, as Darwin intimated, from the Christian's fear of the wrath of an unseen God.[51] Certainly few English sportsmen would have difficulty reading humanlike emotions off the expressions displayed by their dogs. The belief that humans shared comparable emotions and expressions with animals accorded with a common intellectual tradition that can easily be traced back to Aristotle. Yet Darwin's own evolutionary analysis in his *Expression of the Emotions in Man and Animals* (1872) has a peculiar and, for us, an unexpected contour, which can only be understood in light of an unusual theory worked out by one of his contemporaries.

Sir Charles Bell's *Expression: Its Anatomy and Philosophy* (1844) displays a research physician's detailed knowledge of facial anatomy and a devoted humanist's understanding of emotional depiction in art and literature. Bell argued that the smiles and frowns, laughs and sighs, beams and grimaces of the human countenance functioned as a natural language by which one soul communicated with another. Ultimately this repertoire of signs, he asserted, referred back to its divine author, who "has laid the foundation of emotions that point to Him, affections by which we are drawn to Him, and which rest in Him as their object."[52] Thus according to Bell, the expression of the emotions served for communication, human and divine.

Darwin read Bell's book with considerable interest. He focused on the physician's precise descriptions of the structure and operation of facial muscles during the expression of emotions. He denied, however, the theological foundation for emotional expression that Bell divined. But in rejecting Bell's particular conception of the utility of emotional response, he rejected completely all notions of utility for the emotions. Emotional display, to be sure, had an evolutionary history. Darwin's many comparisons of facial patterns in children, adults, the insane, as well as in apes, dogs, and cats—done with the

51. Ibid., 67–68.
52. Charles Bell, *Expression: Its Anatomy and Philosophy*, 3rd ed. (New York: Wells, [1844] 1873), 78.

aid of photography and sketches—showed similarities across ages, sexes, and mental capacities. This kind of comparative evidence bespoke a common origin for emotional expression. Since he could discover no social or communicative function in these emotional reactions, however—unlike contemporary neo-Darwinians—he could not employ his conception of natural selection to give them account. As a consequence, he fell back on his notion that instinctive reactions could derive from practices that had been, by dint of exercise, scored into the heritable substance of human beings. He argued that among our ancestors, if an emotion originally elicited by an appropriate cause produced a certain feeling and consequent expression, then later renewal of the feeling alone could produce the reaction. For example, the turning away and the wrinkled nose of disgust, elicited originally by sight of some repulsive object, might again be displayed as a result of the feeling alone. Darwin called this phenomenon the "principle of serviceable associated habits" and used it to explain, variously, frowning, dejection, smiling, and the like. He formulated two more principles to handle other kinds of expression. The "principle of antithesis" specified that when certain actions were connected with a particular state of mind, an opposite state would tend to elicit an opposite action. For instance, a hostile dog will stand rigid with tail stiff and hair erect, while a docile, happy animal will crouch low with back bent and tail curled. Finally, there was the principle (borrowed from Herbert Spencer), according to which a violent emotion might spill over to adjacent nerve pathways and produce an outward effect—when, for example, great fear caused trembling.[53]

## CONCLUSION

Among the many sources for Darwin's ideas about nature, German Romanticism supplied one of the deeper and more powerful currents. Richard Owen served as one especially important conduit for this tradition. His Goethean morphology and Schellingian archetype theory, suitably reconsidered, formed staples of Darwin's own intellectual repertoire. The doctrine of embryological recapitulation, a fundamental feature of German Romantic biology, became a main supporting pillar of Darwin's general theory.[54] Perhaps the deepest personal source for romantic conceptions of nature came from Humboldt. Darwin modeled his *Researches of the Voyage of the Beagle* on Humboldt's *Personal*

53. Charles Darwin, *The Expression of the Emotions in Man and Animals* (Chicago: University of Chicago Press, [1872] 1965), 28, 28–29.

54. See Richards, *Romantic Conception of Life*, chap. 14, for further discussion of Darwin's relation to the German Romantics.

*Narrative*; Humboldt, that doyen of German science in the first half of the century, returned the compliment by singling out in his book *Kosmos* the merits of the young English adventurer.[55] Humboldt conceived nature as an organism exhibiting interacting parts; Darwin, rejecting the clockwork universe of his English heritage, discovered many ingenious ways of tracing out those organic interactions in the *Origin*. Humboldt's nature had those aesthetic, moral, and creative properties characteristic of the retired Deity, and, as I have tried to show in the previous two chapters and in this one, these are exactly the features exhibited by natural selection. Darwin initially kept the English God unemployed in the background of the *Origin*, where he remained on the dole, ceding the creative work to nature. We usually take the Anglophilic measure of Darwin's ideas from the photo by Julia Cameron, who portrayed Darwin as a sad English prophet (see fig. 5.2). In his youth, however, this fixture of the Victorian establishment sailed to exotic lands, became intoxicated with the sublimity of their environs, and tested his mettle against the forces of man and nature. Like many of the Romantics, he also discovered the moral core of that nature and continually reckoned with it as he constructed his general theory of evolution.

Mind, morals, and emotions occupied Darwin's attention in his early notebooks and found a place even within the *Origin of Species*, which ostensibly avoided the problem of human evolution. His argumentative strategy in the *Descent* and in the *Expression of the Emotions* continued that of the *Origin*. He employed vast amounts of empirical evidence gathered from many different sources and was able to show that, when properly juxtaposed, evolutionary consequences quite naturally fell out. But he did not simply rely on the observations of others. He, of course, made use of his own experience on the *Beagle* voyage, especially his knowledge of tribal life among the Indians of South America and his encounters with the slave trade. Further, he stuffed these books with experiments and mathematical calculations of his own devising. The language of his arguments and experiments did not have the dry, crusty sound of many of the empirical studies from which he drew. His prose had a poetic lilt, and his tropes, such as nature scrutinizing the internal fabric of organisms, allowed the reader to feel the more comfortable presence of a larger power watching over all of life. His metaphors, however, carried a more significant burden. Their evocative surface encased a deep conceptual grammar that structured his thinking about nature so as to represent it as an

55. Alexander von Humboldt, *Kosmos. Entwurf einer physichen Weltbeschreibung*, 5 vols. (Stuttgart: J. G. Cotta'scher Verlag, 1845–62), 2:72.

intelligent, moral agent, one that finally intended "the most exalted object we are capable of conceiving, the production of the higher animals."[56] In this respect, as well, he took his lesson from Humboldt, who supposed that aesthetic judgment might provide an approach complementary to analytic judgment for understanding nature.

Many of Darwin's arguments had the multiply dependent structure of nature herself. He would advance several possible causes to explain the same event, holding those events in a tangled bank of organic relations. Thus, not only did he account for man's big brain by appeal to group selection, but he had the inherited effects of language by which to reinforce his naturalistic theory. He secured human moral character with the interacting forces of community selection, reciprocal altruism, and inculcated habit. The principal force, community selection, along with an evolving intellect, would ensure that human nature might preserve an authentic moral core. As he interpreted his own accomplishment, his theory thus escaped the reproach of grounding human moral capacity in "the base principle of selfishness." Darwin's subtle, artistic effects, along with his voluminous evidence and compelling arguments, have rendered his conclusions powerful even today for the supple of mind.

## APPENDIX: ASSESSMENT OF DARWIN'S MORAL THEORY

Darwin established the framework for a theory of ethics that meets, I believe, both empirical standards of descriptive confirmation and the normative standards that we expect of a theory of morality. The empirical adequacy is measured by anthropological studies of pre-industrial societies, animal studies of altruistic behavior, and general implications from evolutionary theory.[57] The preponderance of evidence indicates that during our evolutionary trajectory, we, like other animals, have acquired altruistic instincts as adaptations to living in social groups. The vehement objections to an evolutionary ethics, however, do not usually rest on the empirical case but on the philosophical case, on providing a normative justification for such an ethics.

56. Darwin, *Origin of Species*, 490. See also chapter 2 in the present volume.

57. The empirical case for the evolution of altruistic systems in animals and men has been argued by a variety of evolutionary researchers and theorists. E. O. Wilson's *Sociobiology* (Cambridge: Harvard University Press, 1975) summarizes a large quantity of the relevant literature; see esp. 106–29 and 547–76. Three other books that offer comprehensive surveys of the empirical evidence are Richard Alexander, *The Biology of Moral Systems* (New York: Aldine De Gruyter, 1987); Lewis Petrinovich, *Human Evolution, Reproduction, and Morality* (New York: Plenum, 1995); and Leonard Katz, ed., *Evolutionary Origins of Morality* (Thorverton, UK: Imprint Academic, 2002).

Thomas Henry Huxley raised the telling difficulty in his lecture "Evolution and Ethics" (1893). Although his animus was directed at his old friend Herbert Spencer, his argument told against Darwin's ethical theory as well. Huxley admitted that our so-called moral impulses, as well as our destructive ones, formed an evolutionary legacy. This fact, however, did not justify any normative judgment: "Cosmic evolution," he admonished, "may teach how the good and the evil tendencies of man may have come about; but, in itself, it is incompetent to furnish any better reason why what we call good is preferable to what we call evil than we had before."[58] Even when confronted with an innate urge to perform a nominally good act, we still must decide whether we *ought* so to perform it. Huxley thus objected that an evolutionary ethics committed what subsequently became known as the "naturalist fallacy" of deriving an "ought" from an "is."[59]

The question may be raised, however, whether the so-called naturalistic fallacy is really a fallacy. There are certain classes of inference we think perfectly valid, whose members are instances of deriving an "ought" from an "is." Darwin himself furnished a relevant example from one of these classes: "The imperious word *ought* seems merely to imply the consciousness of the existence of a persistent instinct, either innate or partly acquired, serving him as guide, though liable to be disobeyed. We hardly use the word *ought* in a metaphorical sense when we say hounds ought to hunt, pointers to point, and retrievers to retrieve their game. If they fail to act, they fail in their duty and act wrongly."[60] Darwin here sanctioned what Kant called an instrumental imperative: given the nature of a thing in a particular surrounding causal matrix, that sort of thing should, or ought, to display certain properties. If moisture, for example, rapidly rises off Lake Michigan when the temperature at ground level is above freezing but quite cold in the upper atmosphere and the wind is blowing due south, we are warranted in saying: "Snow ought to fall on Chicago." This is the sort of prediction in which the "ought" signals that unforeseen causes may interfere, preventing the expected outcome. In the moral situation, if a given individual has evolved to have a particular set of instincts, say other-regarding, altruistic impulses, then in the right circumstances we certainly would be

58. Thomas Henry Huxley, "Evolution and Ethics" (1893), in *Collected Essays*, 9 vols. (New York: D. Appleton, 1896–1902), 9:60.

59. G. E. Moore, in his *Principia Ethica* (Cambridge: Cambridge University Press, 1903), declared the naturalistic fallacy to be identifying a moral good with a natural state. Moore believed the concept of moral good was simple and unanalyzable, and thus not to be defined by any other concept (58). To derive a moral principle from a nonmoral premise, then, would be to commit the fallacy Moore named.

60. Darwin, *Descent of Man*, 1:88.

justified in saying that person *ought* to act in a moral fashion. For example, if John has evolved to have altruistic impulses, then when he sees a friend floundering in deep water just off shore, we rightly expect him to exercise those impulses, that is, we say: "He ought to help Robert." This understanding of the justification of moral judgment is rather like Aristotle's. We make a judgment of moral requirement based on the character of the individual, and more generally, on the fact that the individual is a human with the normal capacities.

It may be objected that this kind of instrumental *ought* does not capture what is usually meant when a moral evaluation is being made.[61] One may wish to urge with Kant that the assessment "John ought to help Robert" is categorical and not dependent on any causal circumstances. Yet if John were tied down, we would not think the moral demand could be made on him. Likewise, I suspect that we would not say "He ought to help Robert" if we knew that John was completely crazy. In other words, causal, factual situations do condition the justification of particular moral evaluations. The most relevant causal context, though, is just that the creature has evolved in a way characteristic of human beings generally.

Few philosophers have written about evolutionary ethics with greater intelligence and more verve than Michael Ruse. In numerous articles he has defended an evolutionary understanding of our moral capacities. Ruse does believe human beings have evolved to have altruistic, or moral, impulses. But these, he thinks, are best understood as biological adaptations that can have nothing other than an empirical justification—nothing that would sanction characterizing these impulses as truly normative. As he puts it: "The point now is that normative ethics is indeed not justified by progress or anything else of a natural kind, for it is not justified in this way by anything." The rules of ethics, he asserts, are like the rules of baseball—you either decide to play or not. If you play, then three strikes make an out, and that's just the fact of the matter, just as, in the morality game, the killing of another human for pleasure is murder and wrong. Actually, though, the decision to play the game is made ultimately by our genes. Our genes, according to Ruse, coerce acquiescence: "here we have a collective illusion of the genes, bringing us all in (except for the morally blind). We need to believe in morality, and so, thanks to our biology, we do believe in morality."[62]

61. Richard Joyce, in his *The Evolution of Morality* (Cambridge: MIT Press, 2005), directs this objection to my analysis (156–60).

62. Michael Ruse, *Evolutionary Naturalism* (London: Routledge, 1995), 249, 250.

If we believe that our genes foist the rules of morality upon us, does this mean we're back with Huxley, who would persist in asking: Yes, but ought we follow these rules nonetheless? Ruse seems to say that there is nothing that compels our following of the rules, except we may recognize that they lead to more harmony in our own lives and allow us to achieve certain desires. But even Ruse would not think that seeking our own pleasure and desires would be anything other than a selfish motive. And if selfishness is the only justification, then for the astute philosopher of evolutionary disposition, all behaviors are warranted, even the most nominally evil—if they satisfy desire.

There is, however, a kind of Kantian twist to the evolutionary analysis. If we have evolved in the way the empirical evidence suggests, then in a cool hour when we contemplate whether we ought to follow an altruistic impulse, we will have no other standard available than the one that leads us to regard altruism as what we ought to sanction. No matter how we contemplate our alternatives—act on the altruistic impulse or act on the selfish—the standard we will employ, if the evolutionary scenario is correct, is that of altruism. Consequently, the fact of our having evolved in a certain way does indeed justify the normative framework we employ. To claim that our genes have coerced and deceived us makes sense only if we are somehow two things—a mind in the prison of the body, which latter has its way with former. But our genes are an essential part of our nature; they help determine who we are. There is no intelligible way a materialist like Ruse can consistently argue the kind of dualism required to sustain his man-in-the-iron-mask notion of human nature. If Darwin is right about our evolutionary history—that is, right about our nature—then our moral standards are indeed justified. They constitute the part of our nature that makes us distinctively human.[63]

63. Much more needs to be said about the justification of an evolutionary ethics. I have tried to say more in Richards, "A Defense of Evolutionary Ethics," *Biology and Philosophy* 1 (1986): 265–93.

CHAPTER FIVE

# The Relation of Spencer's Evolutionary Theory to Darwin's

Our image of Herbert Spencer (1820–1903) is that of a bald, dyspeptic bachelor, spending his days in rooming houses and fussing about government interference with individual liberties (fig. 5.1). Beatrice Webb (1858–1943), who knew him when she was a girl and young woman, recalls for us just this picture. In her diary for 4 January 1885, she writes:

> Royal Academy private view with Herbert Spencer. His criticisms on art dreary, all bound down by the "possible" if not probable. That poor old man would miss me on the whole more than any other mortal. Has real anxiety for my welfare—physical and mental. Told him story of my stopping cart horse in Hyde Park and policeman refusing to come off his beat to hold it. Want of public spirit in passers-by not stopping it before. "Yes, that is another instance of my first principle of government. Directly you get state intervention you cease to have public spirit in individuals; that will be a constantly increasing tendency and the State, like the policeman, will be so bound by red-tape rules that it will frequently leave undone the simplest duties."[1]

Spencer appears a man whose strangled emotions would yet cling to a woman whose philosophy would be completely alien to his own, as Webb's Fabian Socialism turned out to be.

1. Beatrice Webb, *The Diary of Beatrice Webb: Volume One, 1873–1892*, ed. Norman Mackenzie and Jeanne Mackenzie (Cambridge: Harvard University Press, 1982), 127–28.

FIGURE 5.1 Herbert Spencer (1820–1903), in 1890. Photo from David Duncan, *Life and Letters of Herbert Spencer*.

Our image of Darwin is more complex than our image of Spencer. We might think of him nestled in the bosom of his large family, kindly, and just a little sad. The photo of him taken by Julia Cameron (1815–1879) reveals the visage of an Old Testament prophet, though one not fearsome but made wise by contemplating the incessant struggle of life on this earth (fig. 5.2). Ernst Haeckel recalled his first meeting with Darwin in 1866: "He had a Jupiter-like forehead, high and broadly domed, similar to Goethe's, and with deep furrows from the habit of mental work. His eyes were the friendliest and kindest, beshadowed by the roof of a protruding brow."[2] These images have deeply colored our reaction to the ideas of each thinker. The pictures are not false, but they are cropped portraits that tend to distort our reactions to the theories of each. If we examine the major features of their respective constructions of evolution, we may be inclined, as I believe we should be, to recalibrate our antecedent judgments—judgments like those of Ernst Mayr (1905–2005), who in his thousand-page history of biology celebrates Darwin over numerous chapters of superlatives but begrudges only three paragraphs to Spencer, "because his positive contributions [to evolutionary theory] were nil."[3] Mayr's attitude is reflected in most histories of science that discuss evolutionary theory in the nineteenth century. Certainly nothing much of value can be expected from a boardinghouse theorist.

Our contemporary evaluations of the ideas of Spencer and Darwin usually proceed, as Mayr's has, from the perspective of present-day science. Accordingly, Spencer's ship appears to have sunk without a trace, while Darwin's has sailed right into the port of modern biology. Our neo-Darwin perspective, I believe, adds to the distortion worked by popular images of these Victorian gentlemen. During the latter part of Spencer's career, his star certainly achieved considerable magnitude, such that his literary productions began to turn a nice profit. And his contemporaries recognized in his ideas comparable intellectual capital. Alexander Bain (1818–1903) regarded him as "the philosopher of the doctrine of Development, notwithstanding that Darwin has supplied a most important link in the chain."[4] In the historical introduction to the *Origin of Species*, Darwin included Spencer as one of his predecessors; he wrote E. Ray Lankester (1847–1929) that Spencer "will be looked at as by

2. Ernst Haeckel, as quoted by his disciple Wilhelm Bölsche, in *Ernst Haeckel: Ein Lebensbild* (Berlin: Georg Bondi, 1909), 179.

3. Ernst Mayr, *The Growth of Biological Thought* (Cambridge: Harvard University Press, 1982), 386.

4. Alexander Bain to Herbert Spencer (17 November 1863), MS 791, no. 67, Athenaeum Collection of Spencer's Correspondence, University of London Library.

FIGURE 5.2 Charles Darwin in 1875. Photo by Julia Margaret Cameron. (Courtesy of the Smithsonian Institution)

far the greatest living philosopher in England; perhaps equal to any that have lived."[5] Darwin's evaluations of Spencer would alternate between astonishment at the philosopher's cleverness and scorn at his inflated abstractions. Yet, the balance tipped heavily to the positive side. Darwin, along with Thomas Henry Huxley (1825–1895), John Stuart Mill (1806–1873), Charles Babbage (1791–1871), Charles Lyell (1797–1875), Joseph Hooker (1817–1911), Alexander Bain, John Hershel (1792–1871), and a host of other scientists of rather less renown, subscribed to Spencer's program of Synthetic Philosophy, which would issue volumes in biology, psychology, sociology, and morality. These Victorian coryphées redeemed Spencer's intellectual capital with real money. Grant Allen (1848–1899), popular science writer of the late nineteenth century, felt such increasing admiration for Spencer's genius that it finally tumbled forth in poetry:

> Deepest and mightiest of our later seers,
> Spencer, whose piercing glance descried afar
> Down fathomless abysses of dead years
> The formless waste drift into sun or star,
> And through vast wilds of elemental strife
> Tracked out the first faint steps of unconscious life.[6]

We may judge that Spencer got the poet he deserves, but we can hardly doubt that he made a significant mark on his contemporaries.

Spencer's star, to be sure, was slow in rising and always included a reflective glow from Darwin's own. In what follows, I want to take the measure of Spencer's theory along three dimensions, which will allow comparison with essential features of Darwin's conception. These are, first, the origin and character of Spencer's general theory of transmutation; then, more specifically, the causes of species alteration; and finally, the particular case of human mental and moral evolution. In this comparison, I think we will find both some undervalued aspects of Spencer's scheme and some problematic aspects of Darwin's. But this reversal of fortune, if real, produces a historiographic paradox: Why the adulation of Darwin and the denigration of Spencer?

5. Charles Darwin to E. Ray Lankester (15 March 1870), in *Life and Letters of Charles Darwin*, ed. Francis Darwin, 2 vols. (New York: D. Appleton, 1891), 2:301.

6. Grant Allen to Herbert Spencer (10 November 1874), MS 791, no. 102, Athenaeum Collection of Spencer's Correspondence, University of London Library.

## GENERAL EVOLUTIONARY SCHEMES

Both Darwin and Spencer eased into their evolutionary notions in pursuit of their early professions and, indeed, aided by similar intellectual resources. Darwin, of course, sailed away on the *Beagle* in December 1831. He came on board both as companion to the mercurial Captain Robert FitzRoy (1805–1865) and as naturalist, reporting on the geology of areas visited and sending back large numbers of specimens to the British Museum. After the return in October 1836, the cataloging and describing began. In March 1837 he reflected on his experiences in the tranquility of his London study, especially on the Galapagos mockingbirds, which he had originally thought only varieties but finally realized were good species. With his grandfather's own evolutionary theories and those of Lamarck as template, he started formulating his ideas about species descent. The intellectual community had dismissed Erasmus Darwin's (1731–1802) and Jean Baptiste de Lamarck's (1744–1829) speculations as untethered, mere fantasies of a poet and a Frenchman, respectively. Now Charles Darwin's own experience provided an anchor, especially as given weight by the considerations of two works he had carried on the voyage. Alexander von Humboldt's (1769–1859) *Personal Narrative of Travels to the Equinoctial Regions of the New Continent*, which had originally inspired Darwin to undertake the unlikely journey, led him to perceive the interconnectedness of life forms, and encouraged him to appreciate their aesthetic features. Charles Lyell's *Principles of Geology* supplied the vast time scale and biogeographical suggestions for suspecting that Lamarckian transformation theory, which Lyell detailed in volume 2 of his work, might have much more to it than the author allowed. And, of course, in early fall of 1838, Thomas Malthus's (1766–1834) *Essay on the Principle of Population*, with its pregnant notion of population pressure, led Darwin to a "theory by which to work," his device of natural selection.[7]

Spencer's early professional experience lacked the grand sweep of Darwin's.[8] As a civil engineer in his late teens, Spencer had his curiosity piqued by the many fossils he discovered while excavating new passages for the railroads. His reading of Lyell's *Geology* moved him, much as it had Darwin, to consider seriously the Lamarckian hypothesis. Lyell had, in the spirit of the Old Bailey, where he had trained as a barrister, presented a fair case for Lamarck's views,

7. Charles Darwin, *The Autobiography of Charles Darwin*, ed. Nora Barlow (New York: Norton, 1969), 120.

8. I have discussed Spencer's intellectual development in Richards, *Darwin and the Emergence of Evolutionary Theories of Mind and Behavior* (Chicago: University of Chicago Press, 1987), chaps. 6 and 7.

but assumed his subsequent refutation would nullify the Lamarckian theory completely. Lyell was obviously too scrupulous in the former exercise and too hedging in the latter, at least for Spencer. Spencer, though, read few books to the end, so he may simply have missed Lyell's crucial closing arguments. Less significant for Spencer than Darwin, however, were the fundamental biological aspects of development and its environmental setting. Spencer was more interested in human social progress, and that was the consideration that lent the tipping weight to Lamarck's thesis.

Spencer's time with his uncle Thomas Spencer, a curate who had a definite political philosophy, kept him mindful of the possibilities of social development without the aid of government. Poor Laws, Spencer came to believe, were only devious instruments to arrest the need to deal with unjust distribution of the ultimate source of wealth, namely, land. In his first book, *Social Statics*, published at his own expense in 1851, he sounded a call similar to that of his contemporary, Karl Marx. Spencer wrote:

> All arrangements . . . which disguise the evils entailed by the present inequitable relationship of mankind to the soil, postpones the day of rectification. A generous Poor Law is the best means of pacifying an irritated people. Workhouses are used to mitigate the more acute symptoms of social unhealthiness. Parish pay is hush money. Whoever, then, desires the radical cure of national maladies, but especially of this atrophy of one class and the hypertrophy of another, consequent upon unjust land tenure, cannot consistently advocate any kind of compromise.[9]

Only if government would step aside and allow natural development to take its course, Spencer suggested, could society avoid armed insurrection.

In his book *Social Statics* and in his 1852 essay "A Theory of Population," Spencer sketched out what that natural development of society would look like. Similar to Darwin, Spencer employed Malthus's notion of population pressure in a way antithetical to the parson's own dreary conclusions. Malthus imagined that human populations would increase rapidly in favorable times, when food supplies were plentiful, and decline into misery as resources became exhausted. Spencer forecast another possibility: as populations grew, individuals would have to accommodate themselves increasingly to difficult circumstances; habits would have to be developed to articulate men into these

9. Herbert Spencer, *Social Statics: or, The Conditions Essential to Human Happiness Specified and the First of Them Developed* (London: Chapman, 1851), 316.

circumstances; and these habits, as well as the anatomical changes they would induce, would sink into the heritable structure of organisms. Thus individuals would increasingly adapt to the requirements of society and eventually achieve perfect biological accommodation. This was a kind of utopian evolutionism, the goal of which Darwin himself would have acceded to—and, in fact, did, but only with a gaze beclouded with as much doubt as hope.

Spencer, in his essay, mentioned another feature of population pressure that echoes of Spencerian tragedy and Darwinian triumph. He wrote: "It is clear, that by the ceaseless exercise of the faculties needed to contend with them [i.e., the complexities of society], and by the death of all men who fail to contend with them successfully, there is ensured a constant progress towards a higher degree of skill, intelligence, and self-regulation—a better coordination of actions—a more complete life."[10] Thus the principle of natural selection oozed out of Spencer's Malthusian thought, but it immediately dried up. In later years, Spencer would point to this passage as indicating his claim to equitable partnership in authoring the theory of evolution that more and more became associated with Darwin's name.

The final aspect of his reconfiguration of Malthus is unadulterated Spencer. He relied on some very antique ideas ultimately stemming from Hippocratic notions of pangenetic heredity. In ancient medical treatises, connections were made between the production of pangenes from various regions of the body, including the nervous system, and the reproductive organs. The Hippocratics proposed that seeds from all parts of the body, bearing the hereditary material, collected in the brain and slid down the spinal marrow to the generative organs. In the early modern period this ancient view gave rise to the notion that masturbation could cause insanity—a great expenditure of seed would, as it were, melt away the brain. Although Spencer may have been oblivious to the physiological theory behind the wobbly speculations of an ancient medical tradition, he added some loose causal observations of his own to propose an inverse ratio between mental conception and biological conception: the greater the mental complexity of the organism, the fewer the number of biological offspring. Hence, as human society progressed mentally toward a more perfect state, population pressure should decrease. Spencer's prediction of sexual frugality thus tempered Parson Malthus's screed against the possibility of social improvement because of overpopulation. Though Darwin wrote a complimentary letter to Spencer on receiving a copy of the essay on population,

10. [Herbert Spencer], "A Theory of Population, Deduced from the General Law of Animal Fertility," *Westminster and Foreign Quarterly Review* 57 (1852): 500.

he thought that the principles of reproduction Spencer advanced were complete nonsense—after all, he had his own large family as counterevidence.[11] Spencer undoubtedly assumed his bachelor state as proof of his own mental prodigality. Yet Spencer has been shown uncannily correct—greater mental work generally yields fewer biological progeny. The reasons for this, however, are not exactly those he supposed.

Spencer's socialist attitudes lost their vigor with age. By the 1890s, he averred that biological adaptation to the social state must diminish in force as the approach to perfect adaptation increases, so that only in infinite time would the utopia of his youthful radicalism be realized. As his own modest wealth increased, he became considerably less enthusiastic about community ownership of land, finding individual ownership more equitable in the long run.

## NATURAL SELECTION VERSUS FUNCTIONAL ADAPTATIONS AS CAUSES OF EVOLUTION

In his early writing on the development hypothesis, Spencer relied exclusively on habit and the inheritance of consequent anatomical modifications to explain adaptations. But with the publication of the *Origin of Species* in 1859, he came, as he admitted to Darwin, to appreciate the power of natural selection. In his letter acknowledging receipt of a complimentary copy of the *Origin*, he also mentioned to Darwin, lest it be overlooked, that he himself had advanced a similar idea long before, but had confined his considerations to human improvement.[12]

In his book *First Principles* (1860), the initial volume in his series titled Synthetic Philosophy, Spencer relied on the idea of an equilibration between outer environmental circumstances and inner biological conditions in order to explain adaptations. The balancing adjustment of an organism would occur as it adopted new habits to deal with an altered environment. These habits would, in their turn, produce heritable anatomical changes and so realign the organism with its external circumstances.[13] In his *Principles of Biology*, which

11. Charles Darwin to Charles Lyell (25 February 1860), in *Correspondence of Charles Darwin*, ed. Frederick Burkhardt et al., 19 vols. to date (Cambridge: Cambridge University Press, 1985–), 8:109–10: "I have just read his Essay on population, in which he discusses life & publishes such dreadful hypothetical rubbish on the nature of reproduction."

12. Herbert Spencer to Charles Darwin (22 February 1860), ibid., 98–99.

13. Trevor Pearce has shown that it is to Spencer we owe the notion of the relation of a unitary, though articulated, environment (instead of particular circumstances) to an organism as an explanation of adaptations. See Trevor Pearce, "From 'Circumstances' to 'Environment': Herbert Spencer and the Origins

he began issuing in fascicles in 1862, he had to recognize, however, two significant causes of adaptation, what he called "direct equilibration"—the Lamarckian idea—and "indirect equilibration," natural selection, or as he preferred to call it, "survival of the fittest."[14] He admitted that survival of the fittest could account for many traits of plants and the simpler accommodations of animals and men. But he stoutly rejected the suggestion that it could explain more complex coadaptations. He illustrated his argument with the case of the extinct Irish elk. In order for the elk's huge rack of antlers to have evolved, its skull must have thickened, its neck muscles strengthened, its vascular network enlarged, and its nervous connections increased. None of these traits, however, would be of any selective value without all of the others—large neck muscles, for example, would be useless without the great rack of antlers. Yet it would be highly improbable that all of these traits would have simultaneously appeared as spontaneous variations to be selected.[15] Their explanation, according to Spencer, had to be found in the gradual and mutual adjustment of different habits, which would ultimately instill coadapted anatomical attributes. Later, in the 1880s, as the heat streaming from the ultra-Darwinians—such as Alfred Russel Wallace (1823–1813) and August Weismann (1834–1914)—began to be felt, Spencer elaborated his argument based on coadaptation in a large, two-part article titled "The Factors of Organic Evolution," whose aim was to show the insufficiencies of natural selection.[16] (I note in passing that Spencer's is exactly the argument that contemporary advocates of Intelligent Design have used to rejuvenate sclerotic Scientific Creationism; they seem to hope this infusion of monkey glands will unleash a new powerful refutation of Darwinian evolutionary theory.) Darwin himself answered this kind of objection—and Spencer specifically—when he spelled out, in his *Variation of Animals and Plants under Domestication*, how natural selection might operate to produce coadaptations. He reminded his readers that artificial selection could obviously produce the kind of coadaptations that Spencer attributed solely to

---

of the Idea of Organism-Environment Interaction," *Studies in History and Philosophy of Biological and Biomedical Sciences* 41 (2010): 241–52.

14. Spencer used the phrase "survival of the fittest" for the first time in his *Principles of Biology*. He introduced the term in a casual way, suggesting that only later did it occur to him as a felicitous expression. See Herbert Spencer, *Principles of Biology*, 2 vols. (New York: D. Appleton, 1866), 2:53: "natural selection will favour the more upright growing forms: individuals with structures that lift them above the rest, are the fittest for the conditions; and by the continual survival of the fittest, such structures must become established."

15. Ibid., 1:445–57.

16. The articles were drawn together in a small book. See Herbert Spencer, *The Factors of Organic Evolution* (London: Williams and Norgate, 1887).

direct equilibration. After all, multiple, mutual adaptations also go into the selective construction of pouter pigeons and sporting hounds.[17]

Wallace urged Darwin to replace the terms "natural selection" with Spencer's version, "survival of the fittest." He thought Darwin's expression too metaphorical and apt to mislead. As we know, Darwin demurred, saying that his original designation had become enmeshed so tightly within the fabric of the whole theory that it could not be extricated without confusion. He did, though, mention Spencer's expression in the fifth and sixth editions (1869 and 1872) of the *Origin*. I think Darwin was right to reject Spencer's alternative, since these two evolutionists were using completely different conceptions. The difference hinged on the creativity of nature. For Spencer, survival of the fittest meant the elimination of inferior types; it was a negative process. The real creativity of nature, in Spencer's view, stemmed from functional adaptations and coordination through habit, with the inheritance of acquired characters molding the structure of organisms. Moreover, survival of the fittest, Spencer emphatically maintained, did not mean survival of the better or the favored. He urged that "very often that which, humanly speaking, is inferiority, causes the survival. Superiority, whether in size, strength, activity, or sagacity, is, other things equal, at the cost of diminished fertility"—and here he harkened back to his population theory. He continued: "and where the life led by a species does not demand these higher attributes, the species profits by decrease of them, and accompanying increase of fertility. . . . Survival of the better does not cover these cases, though survival of the fittest does."[18] So, for Spencer, survival of the fittest meant, generally speaking, elimination of traits not conducive to greater reproduction, not the selection of favorable attributes and the building up of progressively better adaptations. Survival of the fittest was an entirely negative process. The creativity of evolution, in Spencer's scheme, was left to Lamarckian functional accommodations. For Darwin, however, natural selection was creative and produced better, more progressively advanced creatures. Darwin, thus, too quickly adopted Spencer's phrase "survival of the fittest" as an equivalent to that of "natural selection."

Darwin's conception of the operations of natural selection had its germination in the theory of nature that he embraced during his *Beagle* voyage and that came to invest the *Origin of Species*. While on the voyage, he read and reread the works of Goethe's protégé Alexander von Humboldt, particularly the

17. Charles Darwin, *Variation of Animals and Plants under Domestication*, 2nd ed., 2 vols. (New York: D. Appleton, 1899), 2:327–29 and 327n.

18. Herbert Spencer, "Mr. Martineau on Evolution," in *Recent Discussions in Science, Philosophy, and Morals*, 2nd ed. (New York: D. Appleton, 1882), 339–40.

*Personal Narrative of Travel to the Equinoctial Regions of the New Continent*, an account of the young German's journey to South and Central America in the years 1799 to 1804. Humboldt's understanding of the character of nature both in the large and in the small (i.e., individual creatures) stemmed from his engagement with various members of the early Romantic movement in Jena.[19] Humboldt depicted a nature pregnant with moral and aesthetic values, and governed by archetypal relationships. It was a nature open both to scientific articulation and to artistic intuition, each complementing the other. The ordering of Humboldt's cosmos did not come from a personal Creator but from the fecund and intelligent resources of nature herself. Spinoza, a favorite philosopher of the German Romantics, had epitomized this view with the phrase *Deus sive natura*—God and nature were one. During the *Beagle* voyage, Darwin absorbed this depiction and rendered nature in the account of his journey much after the manner of the German Romantics. He reflected on his debt to Humboldt during his return voyage to England, when he wrote in his diary: "As the force of impression frequently depends on preconceived ideas, I may add that all mine were taken from the vivid descriptions in the Personal Narrative which far exceed in merit anything I have ever read on the subject."[20] In the 1840s, when Darwin was attempting to formulate for himself the character of natural selection, he employed a potent metaphor. He likened the operations of selection to an all-powerful being, one that acted rationally and with forethought, designing adaptations not simply of utility but of aesthetic beauty as well. When this same creature made its appearance in the *Origin of Species* fifteen years later, it had shed some of its garb but none of its deep vitality and moral temper, as I indicated in the second chapter of this volume.

Through means of a literary device, an aesthetic instrument, Darwin infused his conception of nature with "the stamp of far higher workmanship"—higher than any human contrivance could evince—namely, species that bred true, as opposed to the constantly deviating artificial productions of man.[21] Natural selection, in Darwin's image-driven language, patently displayed attributes that Spencer would have denied. Nature did not destroy but rather

19. I have discussed the early German Romantic movement and Humboldt's particular views in Richards, *The Romantic Conception of Life: Science and Philosophy in the Age of Goethe* (Chicago: University of Chicago Press, 2002).

20. Charles Darwin, *Beagle Diary*, ed. R. D. Keynes (Cambridge: Cambridge University Press, 1988), 443 (September 1836). These remarks were reprinted in Charles Darwin, *Journal of Researches into the Geology and Natural History of the Various Countries Visited by H.M.S.* Beagle (London: Henry Coburn, 1839), 604.

21. Charles Darwin, *On the Origin of Species* (London: Murray, 1859), 83–84. See also chapter 2 in the present volume.

creatively directed development in an altruistic and progressive way: "as natural selection works solely by and for the good of each being, all corporeal and mental endowments will tend to progress toward perfection," says Darwin in the *Origin*.[22] Darwin's notion of natural selection as a dynamic, creative force instilling value into nature undoubtedly has had a subtle, even preconscious appeal to the readers of the *Origin*, satisfying a deep need to find some solace in a world from which a creator God had fled. Spencer, by contrast, left his readers with a colder, darker view of the destructive power of nature. Here, then, is Spencer's legacy to our contemporary conception of natural selection, namely, as a destructive force, not acting "solely by and for the good of each being" but rather eliminating most beings and their seed.

## HUMAN MENTAL EVOLUTION

Spencer initially worked out his theory of evolution in light of his utopian socialist vision—a gradual accommodation of human beings to the requirements of social living, so that the greatest amount of intellectual and ethical satisfaction might be achieved. Mental evolution was thus a principal concern right from the beginning of his evolutionary theorizing. The first book of his to achieve some public attention was the *Principles of Psychology*, published in 1855. Spencer had outsized aspirations for this treatise. He predicted that his book would achieve the same intellectual prominence as Newton's *Principia*—at least he so confided this hope to his father.[23] He believed he had resolved a dispute between the followers of Locke and those of Kant—a dispute then at the boil in the exchanges between John Stuart Mill and William Whewell (1794–1866) on the status of universal knowledge claims. The Lockeans maintained that all knowledge was acquired from experience, while the Kantians held that some propositions of universal and necessary modality were innate and determinatively valid. In his *Principles of Psychology*, Spencer argued in Solomonic fashion and came to a conclusion that many philosophers and psychologists today—especially those traveling under the name "evolutionary psychologist"—would endorse. He asserted that certain ubiquitous relationships in the experience of our ancestors concerning space, time, and causality had become impressed on their nervous systems and rendered heritable by dint of constant impregnation. So today those epistemological connections would stand as intrinsic mental structures and serve as the foundation for a

22. Darwin, *Origin of Species*, 489.
23. David Duncan, *Life and Letters of Herbert Spencer*, 2 vols. (New York: D. Appleton, 1908), 1:98.

priori propositions in mathematics and physics. Spencer thus offered an evolutionary Kantianism as the revolutionary account for the foundation of the sciences.

Spencer sent Darwin a copy of his *Principles of Psychology* in early 1856, undoubtedly because he had heard from his friend Huxley that the reclusive naturalist was also working on a theory of descent.[24] Darwin's marginalia indicate he certainly read the book, if without deep penetration. He never mentioned Spencer's work in the early editions of the *Origin* or in the *Descent of Man*, in which latter he revealed his own theories of human mental evolution. Just after the publication of the *Descent*, Spencer wrote his American promoter Edward Youmans (1821–1887) to complain: "As no one says a word in rectification, and as Darwin himself has not indicated the fact that the *Principles of Psychology* was published five years before the *Origin of Species*, I am obliged to gently indicate this myself."[25] The message finally got home to Darwin, and in the last edition of the *Origin*, in 1872, he altered a concluding passage to say: "Psychology will be securely based on the foundation already well laid by Herbert Spencer, that of the necessary acquirement of each mental power and capacity by gradation." Despite Darwin's appraisal of Spencer, he seems to have been little directed by Spencerian ideas, nor was Spencer greatly influenced by Darwinian notions on questions of mental evolution. Each keeping an eye on the other, they nonetheless developed closely parallel conceptions.

As early as the 1840s, Spencer had proposed that the continued development of society and the slow adaptation of its members to the social state would produce "mental and moral and through them, the social perfection of the human race."[26] When he began constructing his Synthetic Philosophy some twenty years later, he retained the conception of the evolutionary process in nature culminating in the moral perfection of human beings: his assessment of cosmic evolution in the initial volume of his system, *First Principles*, and of biological, psychological, and social evolution in subsequent volumes—these all had the purpose of grounding a science of morals in evolutionary processes. He brought his system to a close in 1893 with the publication of his *Principles of Ethics*.

A comparable trajectory can be made out for Darwin. Though his initial thoughts were directed to animal adaptations, he quickly swung to the notion that the highest activity of the human animal—moral behavior—had to

24. Darwin's note of thanks for the book (11 March 1855) is in *Correspondence of Charles Darwin*, 6:56.
25. Herbert Spencer to Edward Youmans (5 June 1871), in *Life and Letters of Herbert Spencer*, 1:197.
26. Herbert Spencer, "Letter VII," *Nonconformist*, 19 October 1842.

be given account by his new theory. His early *M* and *N* notebooks, and his "Old and Useless" notes, kept from 1837 to 1840—all of these contain reading summaries and theorizing about human mental and moral transformation, leavened with recollections of his recent experience of the behavior and mental condition of the South American Indians, the Fuegians, and the Indians of the Pampas. During the despicable effort of the Spanish to exterminate the Indians of Argentina, Darwin detected noble and altruistic behavior exhibited by individuals whom the colonials regarded as little better than animals. And the Indians exhibited moral courage without benefit of the Christian religion. Darwin thought his theory could explain such behavior, and he felt the urgency to do so, lest a crack be left open for the Divinity to be reinstated as a toiling joiner, working piecemeal in the construction of plants and animals.

As Darwin worked out his early theory of moral evolution, he stumbled across a problem that threatened his account not only of human behavior but of his entire theory. This was the difficulty of the social insects—ants, bees, and termites. Soldier bees, for example, would sacrifice their lives for the welfare of the hive, yet since they were neuters, their behavior could not be inherited by their offspring; moreover, even if they could reproduce, such altruistic behavior would have the same effect as if they were neuters—dead bees don't leave progeny. Hence, natural selection could not, it seems, explain the altruistic behavior of such insects. Darwin worried about this potentially crucial objection to his theory right through the late 1850s. Just before the publication of the *Origin*, however, he hit upon the solution: natural selection would operate on the entire hive or community of insects. Hence, those hives that by chance had members exhibiting altruistic behavior would have a selective advantage, and their members, who would include the relatives of the self-sacrificial soldiers, would survive to propagate another day.[27]

Darwin's solution to the problem of the social insects became the model for his explanation of human moral behavior in the *Descent of Man*. The explanation was elegant and one that many of us would still endorse. As I indicated in chapter 4, Darwin contended that among early tribal communities, those that had members who by chance were more cooperative and socially responsive to their fellows would have the advantage over communities that had fewer such members. The cooperative societies would thus be preserved. As the process of community selection continued over long periods of time, new daughter

27. See chapter 4 in this volume for a more detailed account of Darwin's consideration of the social insects.

communities would form, with each generation of members having ever-increasing altruistic dispositions.[28]

Although Darwin focused on human moral acts as that kind of behavior most elegantly explained by his theory of community selection, he found the model to be generalizable. It could also explain growth in human intelligence. A tribe that by chance had a primitive Newton in its midst would profit by adopting his inventions and conceptual notions. This would give the tribe an advantage in competition with other tribes, and so it would be selected, along with that ersatz Newton's relatives.[29] Again, mental traits that might not seem to be greatly advantageous to an individual might yet be selected at the community level and thus continue to advance.

So powerful was Darwin's notion of community selection that Spencer, too, yielded to its attractions, though the philosopher retained the notion that complex traits of organisms—including complex moral behavior—required a theory of direct equilibration for their explanation. In the first part of his *Principles of Ethics*, which was initially published as *Data of Ethics* in 1879, Spencer distinguished two levels of altruism: one kind directed to the family and another to the larger society. He obviously found Darwin's conception of community selection, now narrowed to the family, a fit explanation for altruistic advantage given to children and more remote relatives. He yet held that self-sacrificial behavior operating for the benefit of the larger society could only be explained by gradual accommodation to the social state—his long-standing explanation that depended on direct inheritance of social characters. He perhaps recognized that group selection of unrelated individuals would not yield heritable advantage. Darwin, by contrast, came to believe that group selection per se could occur even if individuals were not related.[30] The problem of group selection still bedevils modern biology.

Darwin's account of human morality largely depended on his theory of community selection, whereas Spencer's still fell back on the inheritance of acquired characters. Yet Darwin could as easily revert to direct inheritance when the situation demanded. This was the case when he focused on the problem of man's big brain. Wallace had pointed out that the human brain had enlarged

28. Charles Darwin, *Descent of Man and Selection in Relation to Sex* (London: Murray, 1871), 1:166.

29. See chapter 4 in this volume.

30. In the final edition of the *Origin of Species* (1872), Darwin invokes group selection quite clearly: "In social animals it [natural selection] will adapt the structure of each individual for the benefit of the community; if the community profits by the selected change." See Charles Darwin, *The Origin of Species: A Variorum Text*, ed. Morse Peckham (Philadelphia: University of Pennsylvania Press, 1959), 172.

capacity, more than what was required for getting along in the world. But if a large brain were not needed for survival, what accounted for the superfluous cerebral matter that most humans carried around? Darwin, after reading some recent German literature, concluded that our acquisition of language molded the brain into more complex patterns, which would become heritable over time. Thus the human brain would grow with the complexity of language.[31] This kind of Darwinian position, though we might cavil about it today, yet reveals a deeper truth that both Spencer and Darwin recognized, namely, that human evolution takes place in society and that social relations become inscribed in the development of the individual.

It has sometimes been suggested that the phrase "social Darwinism"—a phrase that carries a large negative valence—be altered to the more historically correct "social Spencerianism," as if Darwin himself should be exonerated of any application of evolutionary theory to human beings. This suggestion obviously lacks all merit. Neither Darwin nor Spencer thought the human animal exempt from the evolutionary framework and consequent theoretical understanding.

## CONCLUSION

Darwin and Spencer relied on the same devices to explain human mental and moral evolution, that is, natural selection and direct inheritance of acquired relations. Each, however, emphasized that causal account about which each felt most proprietary—certainly no surprise there. No contemporary biologist would, though, be thoroughly satisfied with the theories of either one. Neither Spencer nor Darwin had, by our contemporary lights, a decent notion of heredity. Darwin had no problem, for instance, with natural selection operating on acquired characters, and, of course, his theory of pangenesis was designed to accommodate a Lamarckian kind of inheritance. Yet we are all neo-Darwinians, while none of us would admit to being a neo-Spencerian (though we might charge our enemies with that). Why the denigration of Spencer and the apotheosis of Darwin? Let me conclude with a few suggestions as to the answer to this question.

First, I believe it is the intuitively clear idea of natural selection—at least in its later formulations—that we admire. Ernst Haeckel was ready to regard natural selection as analytically true and thus an immoveable rock upon which to

31. See chapter 8 in this volume for further elaboration of Darwin's use of linguistic considerations.

build evolutionary biology. Of course, Karl Popper (1902–1994) also presumed it as analytically true but drew a different conclusion as to its status in science. We now regard natural selection, under its Spencerian rubric of "survival of the fittest," no longer as an analytic proposition: survival is a criterion of fitness, while fitness itself is a causal condition of survival. But there is more. Darwin's original conception gave natural selection a function—namely, its creative action—to which we are more favorably disposed than to the negative function of elimination that Spencer assigned to it. This, I believe, is a second reason for the prospering of Darwin's fortunes.

One cannot dismiss a related aspect of Darwin's evolutionary views. They *seem* to be based on large and disparate accumulations of empirical evidence. There is, yet, some illusion in this assumption, since the *Origin of Species* makes almost no use of the kind of empirical evidence we today would normally regard as demonstrative, namely, the fossil record. Indeed, Darwin's first German translator, Heinrich Georg Bronn (1800–1862), leveled as a most potent objection to Darwin's theory that he offered only a *possible* scheme of species descent but lacked empirical evidence for the *reality* of species descent.[32] Bronn had a point. The *Origin of Species* is filled with a great variety of stories about how life may have evolved. And so powerful are they that readers have been led simply to accept them as quasi-proofs that life has actually evolved. What Darwin does show is that the kinds of facts with which naturalists would be familiar all hang together in unexpected ways when viewed through the lens of his theory. Spencer's leaden prose could not accomplish the same linguistic magic as Darwin's metaphorical and image-filled writing.

A fourth reason for the ascendancy of Darwinism is that Thomas Henry Huxley and G. E. Moore (1873–1958) indicted Spencer's evolutionary ethics with the charge of committing a great fallacy—the so-called naturalistic fallacy: that because we have, as *a matter of fact*, evolved to regard certain actions as good or bad, we therefore *ought* to regard them as good or bad. Neither Huxley nor Moore appeared to notice that Darwin himself had committed, from their point of view, the same fallacy. As for myself, though, I think it's not a fallacy—but that's irrelevant here.[33]

32. In the translator's epilogue to the German version of the *Origin*, Bronn argued that Darwin had only shown that the kind of transformationism he advocated was possible, but he had not shown it was actual. See H. G. Bronn, "Schlusswort des Übersetzers," in Charles Darwin, *Über die Entstehung der Arten im Thier- und Pflanzen-Reich durch natürliche Züchtung, oder Erhaltung der vervollkommneten Rassen in Kampfe um's Daseyn*, trans. H. G. Bronn and based on the 2nd English ed. (Stuttgart: Schweizerbart'sche Verhandlung und Druckerei, 1860), 495–520.

33. I discuss the logic of evolutionary ethics and the naturalistic fallacy in the second appendix to Richards, *Darwin and the Emergence of Evolutionary Theories of Mind and Behavior.*

A fifth reason for the low estimate of Spencer's program surely has to do with his notions about the liabilities of government interference in natural processes of human development; those notions run counter to most academically liberal sensibilities. Darwin made few preachments about the role of government, especially since his own social position seemed much in harmony with the status quo of his society.

Finally, there are those indelible portraits of the sour bachelor and the prophetic sage. They do work on the imagination.

There is no chance that a space will suddenly be made for Spencer in the pantheon of great scientists. He even failed to find a place on the portico of the Social Science Research Building at the University of Chicago. He was a candidate to have his visage sculpted on the spandrels of the arches, along with Galton, Smith, Gibbon, Bentham, Boas, and Comte. By 1929, however, when the building went up, his time had passed. A historically sensitive reading might remove Spencer from the lower intellectual depths where he now resides, since his impact was felt throughout the nineteenth and early twentieth centuries, often in surreptitious ways. He certainly deserves more than the three paragraphs granted him by Ernst Mayr.

CHAPTER SIX

# Ernst Haeckel's Scientific and Artistic Struggles

Ernst Haeckel was Darwin's foremost champion, not only in Germany but throughout the world (fig. 6.1). In the first decades of the twentieth century, the great historian of biology Erik Nordenskiöld judged that Haeckel's *Natürliche Schöpfungsgeschichte* (12 editions,1868–1920) was "the chief source of the world's knowledge of Darwinism."[1] Haeckel's *Die Welträtsel*, published in 1899, sold over 400,000 copies prior to the First World War and was translated into most of the major languages and several of the more esoteric ones (e.g., Esperanto).[2] Despite his impact on the field of biology—or perhaps because of it—Haeckel provoked a hostile reaction in his own time, especially from the religiously minded; that opposition has been sustained in the present day by those committed to fundamentalist religion.

Haeckel's reputation as researcher, evolutionist, and polemicist brought the kind of fame to his small university in Jena that it had not enjoyed since Goethe administered its affairs a half century earlier. Haeckel drew to Jena the best biologists of the next generation, including the "golden brothers" Oskar (1849–1922) and Richard Hertwig (1850–1937), Wilhelm Roux (1850–1924), and Hans Driesch (1867–1941), all of whom made their mark in science by the end of the century.

Haeckel's students responded not only to his iconoclastic attitudes and aggressive intelligence but also to his unflagging energy and bold creativity.

1. Erik Nordenskiöld, *The History of Biology*, trans. L. B. Eyre (New York: Tudor, 1936), 515.

2. Erika Krauße, "Weg zum Bestseller, Haeckels Werk im Licht der Verlegerkorrespondenz," in *Der Brief als Wissenschaftshistorische Quelle*, ed. Erika Krauße (Berlin: Verlag für Wissenschaft und Bilding, 2005), 145–70.

FIGURE 6.1 Ernst Haeckel (1834–1919), standing, and his assistant Nikolai Miklucho (1846–1888), on the way to the Canary Islands in 1866. Haeckel had just visited Darwin at Downe. (Courtesy of Ernst-Haeckel Haus, Jena)

He introduced into biology many concepts that remain viable today, including the idea that the nucleus of the cell contains the hereditary material, as well as the concepts of phylogeny, ontogeny, and ecology. He was among the first to use the graphic device of the evolutionary tree, and he made it a fixture of biological literature (fig. 6.2). He introduced the idea of the missing link between man and the lower animals, and his speculations led his protégé

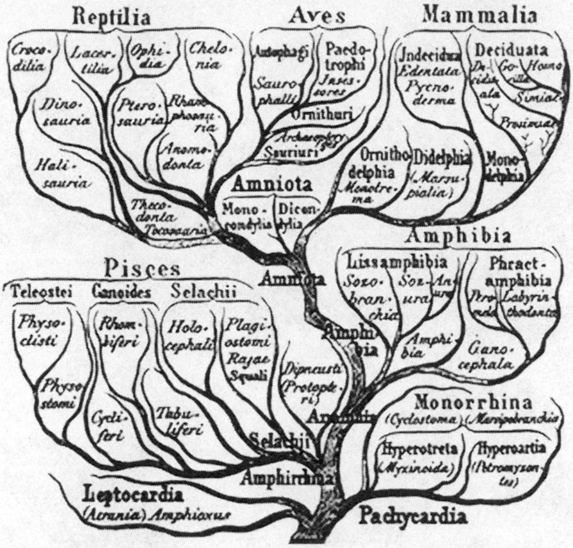


FIGURE 6.2 Stem-tree of the vertebrates. From Ernst Haeckel, *Generelle Morphologie der Organismen* (1866).

Eugene Dubois (1858–1940) to search for its remains in the Dutch East Indies, where he discovered the first *Homo erectus* fossils. Haeckel made central to his evolutionary analyses the biogenetic law, that is, the principle that ontogeny recapitulates phylogeny. The law states that the embryo goes through the same morphological stages in its development as the phylum has gone through in its evolutionary descent. According to this law, the human embryo, for example, begins as a one-celled creature, just as we suppose life began in the sea as a single reproducing cell; then the embryo takes on the form of an invertebrate, then something like a fish, then a mammal, then a primate, and finally a specifically human being. The biogenetic law implies that at the earliest stages of embryogenesis, embryos of a particular order or family ought to be similar in morphology, since their species stemmed from a common ancestor. This is what Haeckel depicted in his many monographs and essays (fig. 6.3).

Haeckel was not only a scientist of extraordinary ability; he was an artist. He supplied the illustrations that would serve for the crafting of woodblocks, copperplate etchings, and lithographs used in the production of his twenty or so technical monographs and innumerable articles. In 1864, he

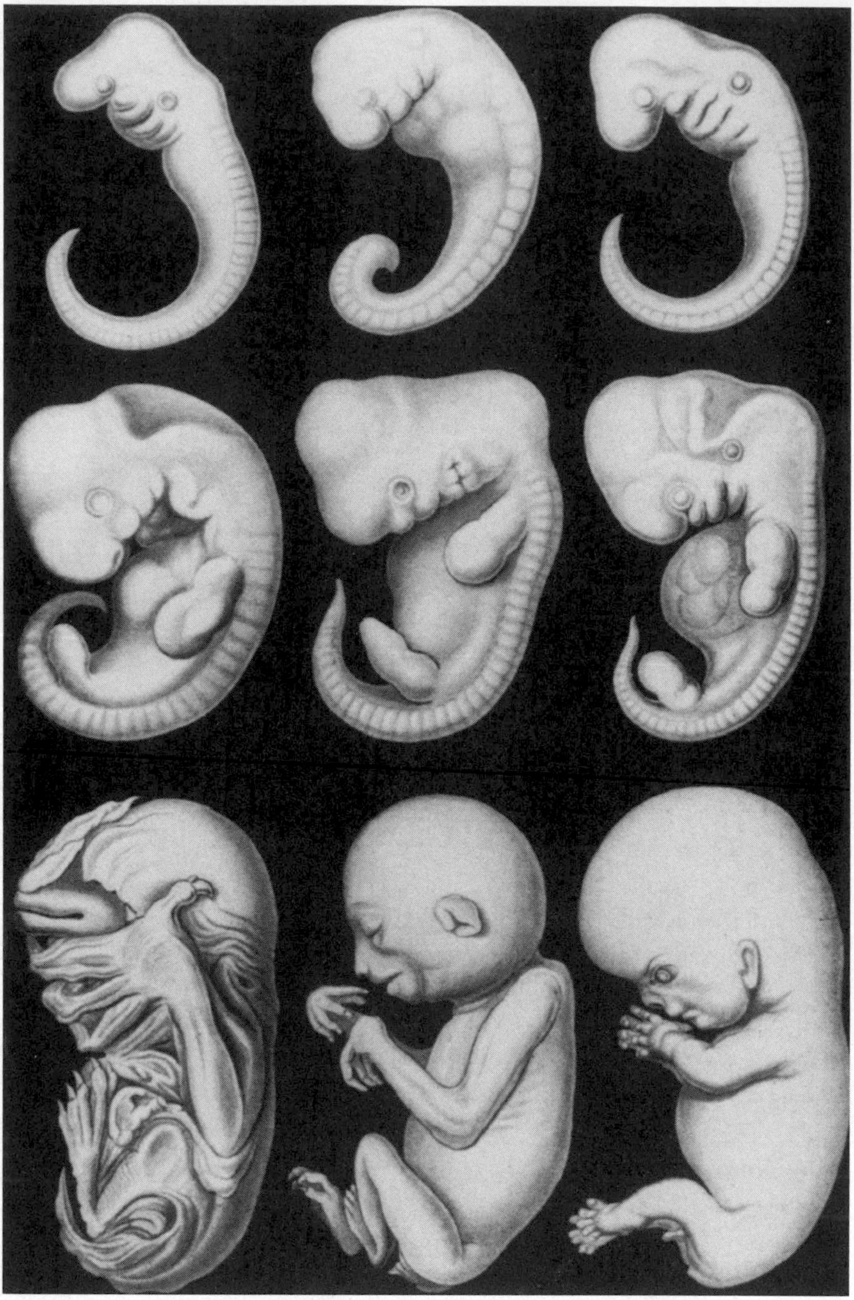

FIGURE 6.3 Illustration of the biogenetic law: comparison of bat, gibbon, and human embryos at three stages of development. From Ernst Haeckel, *Das Menschen Problem* (1907).

sent to Charles Darwin, by way of introduction, his massive work titled *Die Radiolarien* (1862), a seven-pound treatise on creatures no bigger than the head of a pin.[3] The book was the result of his habilitation research in Italy and Sicily. The illustrations in the book, copperplate etchings from drawings by Haeckel, astounded Darwin with their beauty (see pl. 1). When Haeckel would travel to the Italian cities of Rome and Naples, or to the Canary Islands, or Ceylon, or Java—or to the more than thirty other research sites he visited during the half century of his scientific life—he would not only carry sketch pads for depicting the variety of creatures he pulled from the seas but he would also bring canvases to capture the landscapes of the countries visited or the vistas of native life. His artistic impulse flowed as deeply as his scientific impulse.

In this chapter, I focus on two central and closely related aspects of Haeckel's accomplishment: first, the way in which his artistic renderings intersected with his science, giving a distinctive cast to that science and involving him in many disputes; and second, the manner in which both his art and science were driven by an overwhelming tragedy. Let me briefly sketch the course of his life to put these considerations into proper perspective.

## THE TRAJECTORY OF HAECKEL'S LIFE

Ernst Heinrich Philipp August Haeckel (1834–1919) was born into an upper middle-class family.[4] His father Karl was a jurist and a minister in the Prussian Court; his mother Charlotte (née Sethe) came from a family of lawyers. His older brother Karl followed in the family tradition and entered the legal profession. Haeckel, however, inclined toward natural history. Karl Haeckel, with fatherly concern, insisted that his younger son obtain a professional degree. Obediently, the adolescent Haeckel matriculated at the medical school in Würzburg, where he studied with Albert Kölliker (1817–1905) and Rudolf Virchow (1821–1902), two of the most eminent biological and medical researchers of the period. He also worked in the laboratory of the great Berlin zoologist Johannes Müller (1801–1858), with whom he intended to do his habilitation after receiving his medical degree in 1858. Müller's suicide, however, disrupted Haeckel's plans. Shortly thereafter, Carl Gegenbaur (1826–1903) at Jena offered to guide the young student's investigations. Haeckel traveled to southern

3. Ernst Haeckel, *Die Radiolarien. (Rhizopoda radiaria). Eine Monographie*, 2 vols. (Berlin: Georg Reimer, 1862).

4. This biographical section is based on my book, Richards, *The Tragic Sense of Life: Ernst Haeckel and the Struggle over Evolutionary Thought* (Chicago: University of Chicago Press, 2008).

Italy and Sicily for his habilitation research, which yielded the work that initiated his correspondence with Darwin. The two became good friends, with a stream of letters passing between them over the next twenty years. During the course of their friendship, which only terminated with Darwin's death in 1882, the English sage entertained Haeckel three times at his country home in the village of Downe.

Haeckel's monograph on radiolaria—along with Gegenbaur's support—secured a position for him as *extraordinarius* professor at Jena. It also made possible marriage, in 1862, to his cousin Anna Sethe (1835–1864), with whom he was engaged during his Italian sojourn. Their deliriously happy life came to an abrupt end eighteen months later, when on the very day Haeckel was to celebrate his thirtieth birthday, Anna suddenly died of what was likely a burst appendix. Her death completely devastated Haeckel; his family feared he might commit suicide in his desperate grief. Even in his elder years, on the anniversary of Anna's death, he seriously contemplated taking his own life. Her death decisively moved him away from religion and led him to adopt a doctrine that promised less but was demonstrably more reliable—Darwinian theory. He wrote to his parents from Nice, where they sent him to recover:

> The last eight days have passed painfully. The Mediterranean, which I so love, has effected at least a part of the healing cure for which I hoped. I have become much quieter and begin to find myself in an unchanging pain, though I don't know how I shall bear it in the long run. . . . You conclude that man is destined for a higher godlike development, while I hold that from so deficient and contradictory a creation as man, a personal, progressive development after death is not probable, more likely is a progressive development of the species on the whole, as Darwinian theory already has proposed it. . . . Mephisto has it right: "Everything that arises and has value comes to nothing."[5]

While walking along the Mediterranean in a miserable state, Haeckel happened to notice a medusa—that is, a jellyfish—in a tidal pool. That creature with its delicate yellow tendrils reminded him of Anna's golden braids, and in his later publication on medusae, he named it in memory of his wife. A few

5. Haeckel to his parents (21 March 1864), in *Himmelhoch Jauchzend: Erinnerungen und Briefe der Liebe*, ed. Heinrich Schmidt (Dresden: Carl Reissner, 1927), 318–19.

years later, he received from a colleague a specimen he thought even more beautiful, and this would become *Desmonema Annasethe* (see pl. 5).

In 1867, three years after Anna's death, Haeckel remarried. She was the twenty-four-year-old Agnes Huschke (1842–1915). Their marriage was hardly successful, except in the biological sense: they had a son, who became a decent painter, and two daughters, one of whom, like her mother, suffered from the nineteenth-century malady of neurasthenia. Through the years in his oppressive household, Haeckel felt his psychic energy gradually wearing away. This left him falling into the arms of another woman in the late 1890s, a story that also ends in tragedy.

## HAECKEL'S DARWINIAN SCIENCE AND ART

Haeckel's major theoretical work on evolutionary theory—his two-volume *Generelle Morphologie der Organismen* (1866)—failed to excite even the zealots who opposed Darwinian theory, so densely packed were the volumes with the details of systematics and the burdens of newly minted neologisms—Ontogenie, Phylogenie, Oecologie, which would take hold, and Tectologie, Platiden, Amphigonie, which would not. To make evolutionary theory more accessible to his colleagues, he gave a series of popular lectures based on the book (1867–68). He quickly redacted the lectures and added a score of illustrations to produce a wildly successful introduction to Darwinian theory, his *Natürliche Schöpfungsgeschichte* (Natural history of creation,1868). The book went through twelve editions up to the time of Haeckel's death in 1919. In this work, Haeckel made the argument that human beings should be brought under the aegis of Darwinian theory; he made this claim even before the Englishman himself had written on human evolution. Through its several editions, the book became ever more replete with illustrations that reduced the complexities of argument to comprehensible and compelling expressions of the various aspects of the evolutionary process. It was also quite polemical. Haeckel rejected with a sneer the kind of religiously oriented biology that had been the standard prior to Darwin's *Origin of Species*. But the book was only a cool breeze compared to the raging dismissal of religion in his *Welträtsel* (World puzzles, 1899), published at the end of the nineteenth century. This latter work was a phenomenal best seller and was translated into at least thirty languages. A reviewer of the English edition for the *New York Times* encapsulated the book's message with mordant directness: "One of the objects of Dr. Haeckel—it would not be unfair to say the chief object—is to prove that the immortality of the human soul and the existence of a creator, designer, and ruler of the universe are simply impossible. He is not at all an agnostic. Far from it. He knows that there can be no immortality and no God."[6]

In addition to his popular works, Haeckel authored some twenty or so large technical monographs on various marine organisms: sponges, siphonophores, medusae, radiolaria, and other creatures. All of these works included illustrations by his own hand, often reproduced as color lithographs. Haeckel also composed two significant art books: his *Kunstformen der Natur* (Art forms of nature), initially published in ten fascicles of ten plates each from 1899 to 1904, and then in book form (1904), as well as in box form with prints unbound and suitable for framing; and his *Wanderbilder* (Pictures of a wanderer) of 1905, also with unbound prints.[7] The *Kunstformen* reproduced many of the plates from his monographs on marine biology, included new illustrations of more advanced animals, and set them all as artistic pieces. Many of the illustrations were newly colored and replicated in lithograph or in autotype (a monochromatic, nonfading print). The printing work was done by Adolph Giltsch, who was also responsible for reproducing the illustrations of most of Haeckel's many monographs. The *Kunstformen* had a decided impact on the artistic movement of Jugendstil—the German version of art nouveau—which flourished at the beginning of the twentieth century.[8] The *Wanderbilder* gathered together landscapes that Haeckel painted on his two trips to the tropics, one in 1881–82, when he traveled to Ceylon (now Sri Lanka), and the other in 1900–1901, when he journeyed to Sumatra and Java. The technical aspects of printing these landscapes required another hand; it was undertaken by the natural history publishing house of Eugen Koehler and his son Woldemar Koehler. Although the plates included photographs of some scenes, especially those of individuals in native dress, the purpose was to induce in the reader a deeper feeling for nature, which Haeckel believed could only be inspired by vivid illustrations based on paintings in oil or watercolor (for example, pl. 2). Photography simply could not produce the desired effect.

Haeckel admitted that he was "no accomplished artist, but only an enthusiastic dilettante whose moderate talent, through extensive practice and heartfelt dedication, has been directed usefully to nature."[9] This deflationary evaluation

6. "A Little Riddle of the Universe," *New York Times*, 27 July 1901.

7. Ernst Haeckel, *Kunstformen der Natur* (Leipzig: Bibliographisches Institut, 1904); Ernst Haeckel, *Wanderbilder: Nach eigenen Aquarellen und Oelgemälden* (Gera-Untermhaus: W. Kochler, 1905).

8. See Christoph Kockerbeck, *Ernst Haeckel's "Kunstformen der Natur" und ihr Einfluß auf die deutsche Bildende Kunst der Jahrhundertwende* (Frankfurt: Peter Lang, 1986).

9. Haeckel, *Wanderbilder*, p. 3 of the unnumbered pages.

belies his aesthetic talent, honed by study and unremitting effort. The scene in plate 2, for example, evokes Kant's notion of the sublime—a feeling of individual insignificance in view of the power of nature, yet with a recognition of human mental power that rises above nature. It is also reminiscent of Caspar David Friedrich's famous *Der Wanderer über dem Nebelmeer* (Wanderer above the sea fog, 1818).

## AESTHETICS AND SCIENCE

Haeckel's artistic efforts and scientific practice were intimately connected along several dimensions. One might first consider the supposed stylized character of his scientific illustrations. Stephen Jay Gould, no friend of his predecessor, maintained that Haeckel made his drawings too symmetrical, too artificially regular, and thus they could not represent the real character of the organisms depicted. Gould had particularly in mind Haeckel's illustrations of radiolaria, as, for example, those pictured in plate 1. More recently, Lorraine Daston and Peter Galison in their book *Objectivity* have leveled a comparable charge, suggesting that Haeckel remained mired in an older tradition, while more empirically inclined naturalists had taken up the camera to render nature with photographic realism and precision.[10] Peter Bowler has argued that Haeckel's artistic representations reveal his non-Darwinian approach. He contends that Darwin emphasized the variability of organisms, the very material of evolutionary adaptation and development, while Haeckel showed no interest in variable traits.[11] I believe these criticisms are unfounded and neglect the intended purpose of Haeckel's science and his art.

Haeckel's depictions of radiolaria do show them as quite symmetrical, because as a matter of fact they are—notoriously so (fig. 6.4). Haeckel's intention in constructing his atlas of radiolaria—as well as the many other atlases accompanying his volumes on the systematic description of medusae, siphonophores, sponges, and other creatures—was to provide a *standard* representation of a given species. Had he included a depiction of a particular individual deviating from the species norm—instead of one exhibiting the essential structure of the species—the illustration would be defective for the purposes of identifying individuals of a species. Moreover, Haeckel understood quite well the advantages of the watercolor or oil painting over the photograph:

10. Lorraine Daston and Peter Galison, *Objectivity* (New York: Zone Books, 2007), 194–95.

11. Peter Bowler, *The Non-Darwinian Revolution: Reinterpreting a Historical Myth* (Baltimore: Johns Hopkins University Press, 1988), 83.

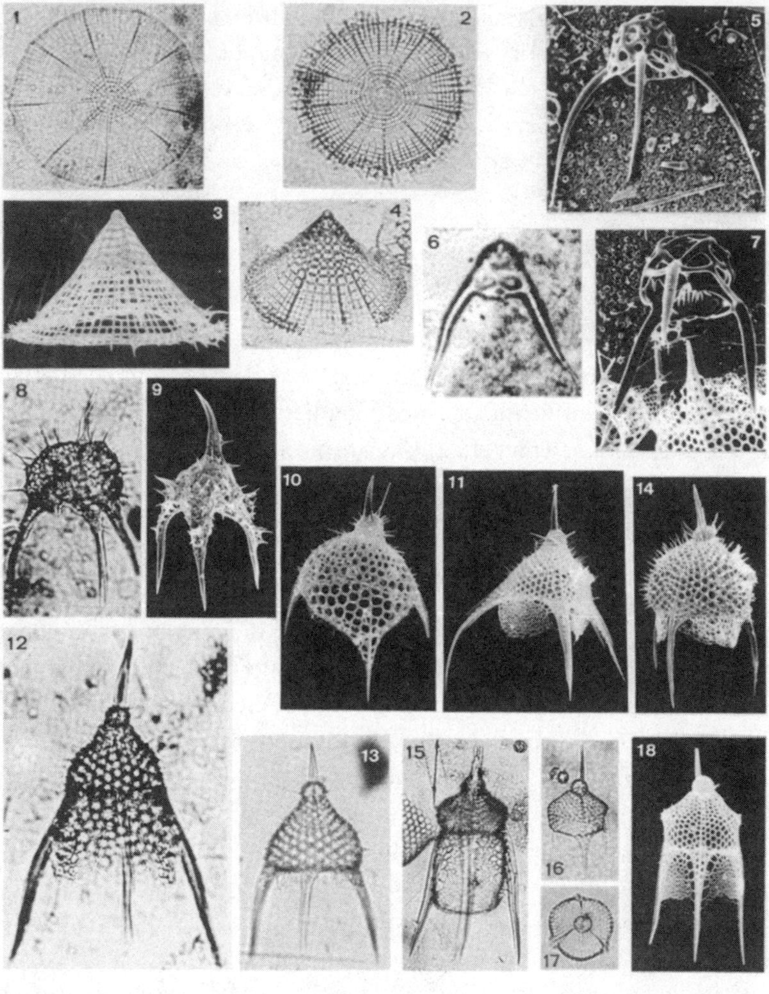


FIGURE 6.4 Micrographs of the subfamilies Plectopyramidinae and Eucyrtidinae. From Kozo Takahashi and Susumu Honio, *Radiolaria* (1991).

I have been convinced that colored images (even a mediocre production) are much more valuable for a vivid intuitive awareness of nature than the photograph or the simple black and white illustration. Indeed, a crude color sketch (if it conveys the landscape in a vivid fashion) has a deeper and more stimulating effect than the best black and white illustration or photographic representation. This distinction lies not only in the

> effect of color itself—since different individuals are sensitive in different measures—but also because the painter—as thoughtful artist—reproduces in his subjective image the conceptually articulated character of the landscape and emphasizes its essential features. The objective image of the photograph, by contrast, reproduces equally all parts of the view, the interesting and the mundane, the essential and the inessential. Thus the colored photograph, if it should be brought to perfection, will indeed never be able to replace the individually conceived and deeply felt image of the painter.[12]

Haeckel understood that when depicting botanical objects, as well as birds, fish, hydrozoa, and most other animals, the color of the subjects was crucial. Of course, color photography, which Haeckel presciently foresaw, would not be perfected until the 1930s. During the nineteenth century, the mode of color reproduction was the etched copperplate or the lithograph, both of which depended on the artist's illustration. Further, the accidental and unrepresentative aspects of creatures, as opposed to their essential features, had to be excluded. Many of the specimens that Haeckel had at his disposal—and would render into striking images, careful to get color and essential features exact—were damaged or defective in some way. They had to be rectified through the experience of the naturalist and the imagination of the artist. For example, a medusa that Haeckel named after his first wife—*Desmonema Annasethe*—originally came to him as a compressed and crumpled brown mass. It was sent to him preserved in spirits of wine and shipped in a soldered tin by his cousin in Africa, the linguist Wilhelm Bleek (1827–1875). Bleek, significantly, was also the cousin of Anna Sethe. Haeckel's initial illustration of this organism (pl. 3), while structurally correct, lacked the vivid colors of the original, qualities it would later acquire in Haeckel's inspired hands. I will come back to this image later in the chapter.

A final reason why photography would not and could not substitute for the artist's brush has to do with light, something Haeckel understood well. While in the highlands of Java in 1900, he meditated on the subject of light and the disadvantage of photography in dealing with its difficulties. He wrote:

> In the colorful confusion produced by the mass of tangled plants, the eye vainly seeks a resting place. Either the light is reduced and distorts the thousand crisscrossed branches, twigs, and leaf surfaces . . . or the light of

12. Haeckel, *Wanderbilder*, p. 3 of the unnumbered pages.

> the overhead sun ... produces on the mirrored surface of the leather-like leaves thousands of glancing reflections and harsh lights, which allow no unified impression to be gathered. In the depths of the primitive forest, the various complexes of light are extraordinary and cannot be simply reproduced by means of photographs.... A good landscape painter—especially when he possesses botanical knowledge, is able in a larger oil painting to place before the eye of the viewer the fantastic magical world of the primeval forest in a realistic way.[13]

The painter's hand can control light, filter it precisely, so that shadows do not obscure nor glare wash out features essential to the understanding of organisms. This aspect of the illustrator's technique was brought home to me by one of my students, who is a biological illustrator. On examining one of her extremely precise—almost photographic—pencil drawings of a vertebrate lower jaw, I asked about the direction from which the light was coming. She quickly said, "we don't worry about that." She indicated that if the direction of light were realistically portrayed, some structures of the bone would be hidden in shadow, while artful shading could not be used to emphasize structures. Haeckel knew this as well.

Haeckel intended to represent not only the essential structural features of radiolaria but also their beauty, which he was able to portray through the use of color and the balanced arrangement of creatures in his atlas plates. Haeckel had been convinced by his mentors Goethe and Alexander von Humboldt that to depict the wonders of nature accurately was not only to discover "the laws of their origin and evolution but also to press into the secret parts of their beauty by sketching and painting."[14] Alexander von Humboldt's *Kosmos* was predicated on this aspect of the naturalist's representations of nature. Yet both Humboldt and Haeckel had an even more radical intention—they wished the observer of their volumes to have an experience comparable to that of the naturalist who first encountered the seductive displays of nature.[15] As Haeckel expressed his intent in *Kunstformen der Natur*: "Nature generates from her womb an inexhaustible cornucopia of wonderful forms, the beauty and variety of which far exceed the crafted art forms produced by human beings." But nature's wondrous structures often lay hidden in the jungles of tropical lands or in the depths of oceans beyond the view of the ordinary reader. By his artistic

13. Ernst Haeckel, *Aus Insulinde: Malayische Reisebriefe* (Bonn: Emil Strauss, 1901), 106–8.

14. Ernst Haeckel, "Vorwort," in *Kunstformen der Natur*, p. 1 of the unnumbered pages.

15. See Alexander von Humboldt, *Kosmos: Entwurf einer physischen Weltbeschreibung*, 5 vols. (Stuttgart: Cotta'scher Verlag, 1845–58), 2:73.

efforts Haeckel sought to "bring those forms into the light and to make them accessible to the greater circle of the friends of art and nature."[16] To accomplish this, the artist-naturalist had to create depictions that would give the reader a partial experience of nature's extraordinary beauty; the naturalist had to allow the reader to share the experience he once had of such extraordinary sights.

Haeckel's conviction about the astounding structures of life hidden from ordinary view was shared by René Binet (1866–1911), the chief architect of the Paris Exhibition of 1900. Binet thought such extraordinary forms should be displayed as a main attraction of the fair. To that end, he used Haeckel's work on radiolaria as motif for the various exhibits, including the entranceway to the fair, the Porte Monumentale (fig. 6.5).

Even if the one-celled radiolaria in fact show a deep symmetry, what about the metazoa, the many-celled creatures that Haeckel also portrayed? Perhaps here the objection might well be erected as a warning about the creative channels of Haeckel's art. His depictions of metazoans are symmetrical and idealized. The individual creatures that Haeckel pulled up from the sea would have lacked the perfection of form exemplified by his illustrations. Take, for example, the beautiful *Physophora magnifica*, flanked by juvenile specimens, that graced his prize-winning monograph on siphonophores (pl. 4).[17] It is obvious that at one level what Haeckel portrayed was not an individual carrying all the marks of particularity—with deflated bells or missing organs—but an ideal, an archetype of the species. While Gould's protest that Haeckel's images were too symmetrical fails in regard to the simple radiolaria, it might well be appropriate in regard to more advanced creatures, like the siphonophores.

To understand Haeckel's artistic and scientific justification of his practice, one must consider the assumptions and principles that guided his hand—and still guide the hands of biological illustrators today. These assumptions and principles, in Haeckel's case, had three sources: first, the morphological tradition in which he was schooled; second, what he came to understand as the object of biological and, indeed, artistic comprehension; and, finally, his deeper evolutionary and metaphysical convictions.

First, then, there is the Goethean morphological tradition. Haeckel had been enamored of Goethe since his youth—and that passion did not wane in his later years. He wooed both Anna Sethe and later Frida von Uslar-Gleichen (1864–1903) with Goethe's poetry. And it was a Goethean morphology of which he was persuaded. Goethe held a version of Spinoza's doctrine of

16. Haeckel, "Vorwort," in *Kunstformen der Natur*, p. 1 of the unnumbered pages.
17. Ernst Haeckel, *Zur Entwicklungsgeschichte der Siphonophoren* (Utrecht: C. van der Post, Jr., 1869).

FIGURE 6.5 René Binet's Porte Monumentale at the Paris Exhibition of 1900. From the author's collection.

*adequate ideas*, that is, the notion that within nature, which Spinoza identified with God, real ideas were to be found, counterparts of material individuals. These ideas, as Goethe construed them, were generative; they were responsible for their material manifestations. In Goethe's view, both scientist and artist had to understand these ideas—or archetypes, as they became known—in order to comprehend natural creatures in a scientific way and to render them aesthetically in artistic productions. Thus, in a given instance, the same archetypal principles would serve the scientist and artist in a complementary pursuit. For Haeckel, then, what he conveyed to his reader analytically in precise description might also be rendered intuitively in an illustration that would reveal the same underlying archetype.

Haeckel's more metaphysical considerations of the Goethean archetype became transformed into a historical scenario after he read Darwin and became convinced that what earlier morphologist spoke of as the archetype could now be understood as the genealogically derived structure of the species (or the Bauplan of the ancestor in the case of the phylum). Thus Goethe's archetype became historicized in Darwinian science, and the unity of type exhibited by various species (e.g., vertebrates) could be traced not to an abstract metaphysical idea but to a common ancestor of those species. Yet Haeckel retained the Goethean conception that the proper object of biological investigation was the archetypal structure of a species, which could now be traced back in evolutionary history to the common ancestor of that species and of those closely related to it. Hence, the subject of his inquiries—the second point I mentioned earlier—was not this particular medusa but the underlying structure that united it with others of its species and ultimately with the ancestor that established the phylum.

Haeckel's science did not abandon a metaphysical foundation, though it had changed after the infusion of a Darwinian historical dynamic. Under the new Darwinian dispensation, however, Haeckel did not deny the reality of the species type and its own more fundamental structure, the phylogenetic archetype. These were indeed real aspects of nature as embodied in particular individuals. There was, though, another kind of metaphysical assumption to which Haeckel's biology gave expression, and it concerned the death of his first wife. This is the third point I wish to consider. That death marked a radical religious and philosophical turning point in his life. As the letter to his parents indicates, he abandoned orthodox religion and replaced it with Darwinian theory. A year after the death of his wife, Haeckel began what would become a two-volume, highly theoretical application of evolutionary theory to morphology. He set a feverish pace, and within fourteen months his thousand-page *Generelle Morphologie der Organismen* was published. This work constituted Haeckel's fundamental position on Darwinian theory and its application to all of life. The last chapter of the book took a sharp metaphysical turn. He followed Goethe and Spinoza in identifying God with nature: *Deus sive natura*. And while Haeckel as scientist recognized that all individuals were mortal, the romantic Haeckel presumed that nature preserved all of life in her bosom. He captured this attitude in the epigram from Goethe that he used as preface to his book: "There is in nature an eternal life, becoming, and movement. She alters herself eternally, and is never still. She has no conception of stasis, and can only curse it. She is strong, her step is measured, her laws unalterable. She has thought and constantly reflects—but not as a human being, but as nature. She appears to everyone in a particular form. She hides herself in a thousand names and terms, and is always the same."[18]

18. Ernst Haeckel, *Generelle Morphologie der Organismen*, 2 vols. (Berlin: Georg Reimer, 1866), 1:iii.

Haeckel seems to have felt that Anna had returned to nature and retained a presence therein. Toward the end of his life, when he produced the *Kunstformen der Natur*, that original, crumpled, brown creature *Desmonema Annasethe* was resurrected into the beautifully transformed medusa that is now emblematic of Haeckel's accomplishments as an artist (pl. 5). In the *Kunstformen*, he remarked: "The species name of this extraordinary Discomedusa—one of the loveliest and most interesting of all the medusa—immortalizes the memory of Anna Sethe, the highly gifted, extremely sensitive wife of the author of this work, to whom he owes the happiest years of his life."[19] He wrote these tributes while still married to his second and apparently forgettable wife Agnes. The creature that appears in the *Kunstformen* has become more beautiful, certainly more beautiful than the brown, compressed exemplar he had received from his cousin. Moreover, the composition is artfully balanced, with *Annasethe* flanked by two other species. The *Chrysaora mediterranea* at the lower right is a venomously armed companion to *Annasethe*. In nature it is about four times the size of *Annasethe*. *Floscula Promethea*, in the upper left, is only a quarter of the size of *Annasethe*. Haeckel adjusted the dimensions of each of the flanking medusae to complement the magnificent creature at the center. Haeckel's first wife grew in memory more beautiful and significant over his lifetime. For Haeckel, love fled and hid her face among sea creatures.

19. Haeckel, *Kunstformen der Natur*, text to Tafel 8.

CHAPTER SEVEN

# Haeckel's Embryos

## *Fraud Not Proven*

In a *Science* magazine article published in 1997, "Haeckel's Embryos: Fraud Rediscovered," Darwin's champion in Germany, Ernst Haeckel, was accused of having intentionally misrepresented embryological development. The article reported that the work of Michael Richardson and his colleagues demonstrated this malfeasance through a comparison of Haeckel's illustrations of early-stage embryos with photographs of the same species at a comparable stage (see fig. 7.1). The photos showed embryos of various species that differed among themselves and certainly from Haeckel's images. The differences were striking and the implication obvious: fraudulent misrepresentation. Richardson, as quoted in the article, affirmed the charge: " 'It looks like it's turning out to be one of the most famous fakes in biology.' "[1] The popular press immediately picked up the story, running it under such headlines as "An Embryonic Liar."[2] It was not long thereafter that creationists and advocates of Intelligent Design ignited thousands of websites in an electronic auto-da-fé wherein Haeckel's reputation and that of Darwinian theory generally were sacrificed to appease an angry God.[3] It had long been assumed that Haeckel's racist construction of human evolution had contributed to the work of the Nazis, and now the photographic evidence seemed to confirm his meretricious

1. Elizabeth Pennisi, "Haeckel's Embryos: Fraud Rediscovered," *Science* 277 (1997): 1435.
2. Nigel Hawkes, "An Embryonic Liar," *Times* (London), 11 August 1997, 14.
3. See also the use made of the work of Richardson et al. by the creationist Jonathan Wells, in Wells, *Icons of Evolution* (Washington, DC: Regnery, 2000), 81–109; and in Wells, *The Politically Incorrect Guide to Darwinism and Intelligent Design* (Washington, DC: Regnery, 2006), 27–29.

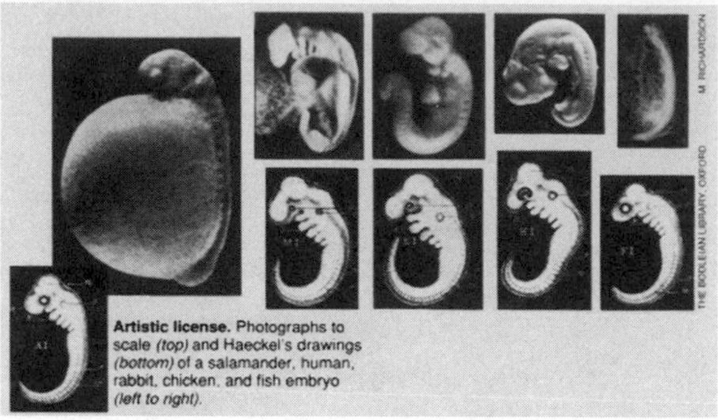

M. RICHARDSON
THE BODLEIAN LIBRARY, OXFORD
**Artistic license.** Photographs to scale *(top)* and Haeckel's drawings *(bottom)* of a salamander, human, rabbit, chicken, and fish embryo *(left to right).*


FIGURE 7.1 Illustration of embryos. From Elizabeth Pennisi, "Haeckel's Embryos: Fraud Rediscovered," *Science* (1997).

character. Many reputable biologists quickly accepted the conclusion of the *Science* article but then sought to distance Haeckel's version of evolution from that of Darwin. Stephen Jay Gould, for example, thought the indictment justified Louis Agassiz's judgment of Haeckel: "*Abscheulich!* (Atrocious!)."[4] Ever since the appearance of his book *Ontogeny and Phylogeny* (1977), Gould had been trying to distinguish Haeckel's evolutionary views from Darwin's—especially concerning the idea that the development of a given embryo morphologically recapitulated the evolutionary history of its phylum. Richardson's evidence gave dramatic support for Gould's many efforts to discredit Haeckel.[5] The historical and biological evidence, however, shows the charge against Haeckel to be logically mischievous, historically vacuous, and founded on highly misleading photography.

*Science* based its report on an article by Richardson and his colleagues in the journal *Anatomy and Embryology* (1997).[6] These scientists argued that vertebrates did not go through an early embryological stage (the so-called phylo-

4. Stephen Jay Gould, "*Abscheulich!* (Atrocious!) Haeckel's Distortions Did Not Help Darwin," *Natural History* 109, no. 2 (2000): 42–49.

5. See, for example, the following works by Stephen Jay Gould: *Ever Since Darwin* (New York: Norton, 1977), 215–17; *The Panda's Thumb* (New York: Norton, 1980), 237–41, 346–47; *The Flamingo's Smile* (New York: Norton, 1985), 90, 412–13; *Wonderful Life: The Burgess Shale and the Nature of History* (New York: Norton, 1989), 263–67; *The Hedgehog, the Fox, and the Magister's Pox* (New York: Harmony Books, 2003), 157–62.

6. Michael Richardson, J. Hanken, M. L. Gooneratne, C. Pieau, A. Raynaud, L. Selwood, and G. M. Wright, "There Is No Highly Conserved Embryonic State in the Vertebrates: Implications for Current Theories of Evolution and Development," *Anatomy and Embryology* 196 (1997): 91–106.

typic stage) in which different species were supposed to be morphologically quite similar, although this had been the conviction of many embryologists of the past and the present. Richardson and his assistants maintained that not only did Haeckel's images misrepresent the actual state of embryos at early stages but so did the illustrations of Wilhelm His, perhaps the most famous embryologist of his day and Haeckel's bitter enemy. His, they contended, also exaggerated the similarities of embryos at early stages and ignored their differences. The main point of the article by Richardson and his colleagues, however, was to show that embryologists in the late twentieth century did little better. The authors, though, accused no one of fraud. The charge of fraud was made in the *Science* article, and then only of Haeckel. Parity of reasoning should logically have required another conclusion: if the indictment of fraud should be made against Haeckel because of too-similar images, then it ought to be brought also against His and the many modern embryologists whom Richardson and his colleagues cited, since they too supposed a phylotypic stage of embryos.[7] Actually, these recent embryologists ought to have been judged more culpable, given the increase of knowledge, refinement of standards, and perfection of instrumentation during the last 125 years.

Richardson and his colleagues chose to compare their photographs with images taken from Haeckel's *Anthropogenie oder Entwickelungsgeschichte des Menschen* (Anthropogenie or the developmental history of man, 1874),[8] one of Haeckel's popular defenses of evolutionary theory. The book grew out of a series of lectures he gave to a general audience in 1873, which he then quickly redacted from stenographic notes taken by two of his students. Haeckel's lectures and his volume were replete with many illustrations by his own hand, including the comparative illustration supporting the recapitulation hypothesis (fig. 7.2). It was from this latter group that Richardson and his colleagues selected images of embryos for comparison with their photographs.

There are several matters of historical importance that one must keep in mind when judging the veracity of Haeckel's work. First, his lectures were

7. Richardson et al. cite the following modern embryologists as believing in a phylotypic stage in which vertebrate embryos very closely resemble one another: H. Butler and B. Juurlink, *An Atlas for Staging Mammalian and Chick Embryos* (Boca Raton, FL: CRC Press, 1987); L. Wolpert, *The Triumph of the Embryo* (Oxford: Oxford University Press, 1991); J. Slack, P. Holland, and C. Graham. "The Zootype and the Phylotypic Stage," *Nature* (1993): 361, 490–92; B. Alberts, D. Bray, J. Lewis, M. Raff, K. Roberts, and J. Watson, *Molecular Biology of the Cell*, 3rd ed. (New York: Garland, 1994); and P. Collins, "Embryology and Development," in *Gray's Anatomy*, 38th ed., ed. P. Collins (London: Churchill Livingstone, 1995), 91–341.

8. Ernst Haeckel, *Anthropogenie oder Entwickelungsgeschichte des Menschen* (Leipzig: Engelmann, 1874).

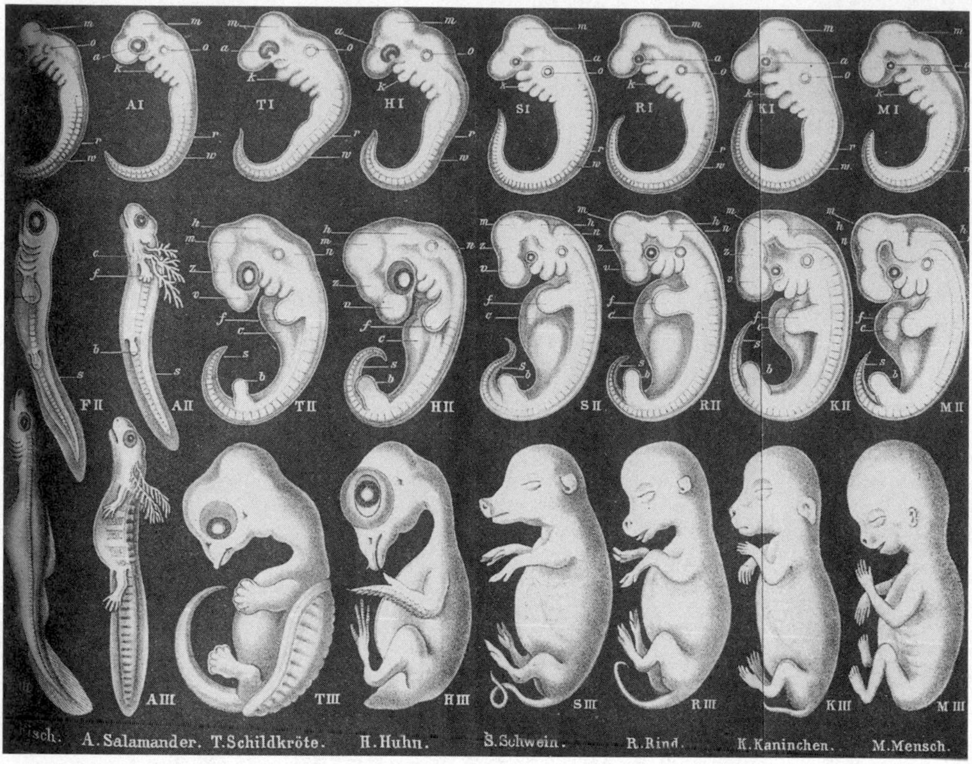


FIGURE 7.2 Illustration of the biogenetic law. From Ernst Haeckel, *Anthropogenie* (1874).

meant for a popular audience, and thus some didactic license would have been permitted. Second, Haeckel was a marine biologist, not a vertebrate biologist, though highly skilled in the latter field. Consequently he borrowed and adapted many of his illustrations, with acknowledgment, from experts in vertebrate biology. From our perspective, these images are a bit crude. If one compares Haeckel's images of embryos at the intermediate stage with those used by Darwin in the *Descent of Man* (fig. 7.3), one can appreciate the schematic character of images typical of the time. Indeed, Darwin acknowledged that he borrowed his images from two of the same sources as did Haeckel.[9]

9. Charles Darwin, *The Descent of Man and Selection in Relation to Sex*, 2 vols. (London: Murray, 1871), 1:16.

CHAP. I. EMBRYONIC DEVELOPMENT. 15

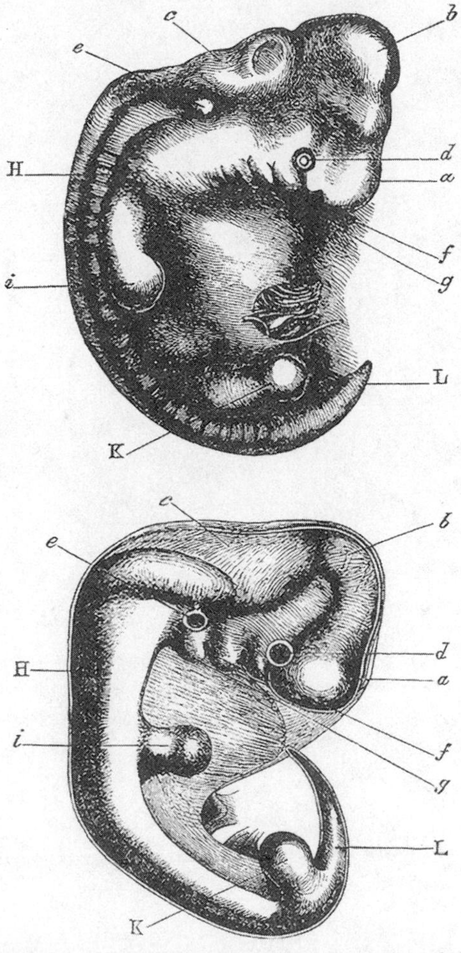


Fig. 1. Upper figure human embryo, from Ecker. Lower figure that of a dog, from Bischoff.

FIGURE 7.3 Illustration showing similarities between human and dog embryos. From Charles Darwin, *Descent of Man* (1871).

Since Darwin also attempted to drive home the similarities of vertebrate embryos, perhaps not even he should escape condemnation. Third, in the *Science* article, Richardson is quoted as suggesting that Haeckel "fudged the scale" of the embryos, even though there was a tenfold difference among them. Haeckel, however, explicitly stated in the caption to his illustration that he had

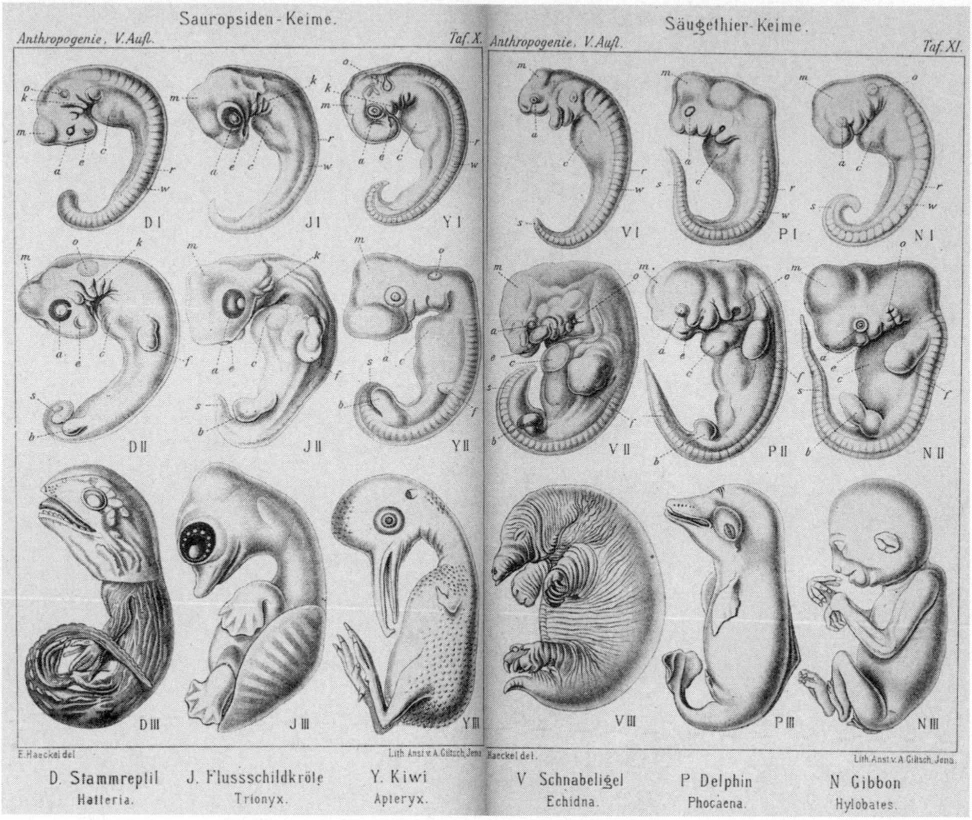


FIGURE 7.4 Illustrations of the biogenetic law. From Ernst Haeckel, *Anthropogenie*, 5th ed. (1905).

reduced all of the images to the same size to facilitate structural comparisons.[10] Finally, Richardson and his colleagues selected images from the first edition of Haeckel's *Anthropogenie*, which was hastily drawn together from his lectures. The book, though, went through five further editions. With each new edition the text grew fatter as Haeckel deployed more evidence, and the illustration in question expanded the comparison from eight species of embryo to twenty (fig. 7.4) by the fifth edition (1905). Moreover, the images grew ever more refined at all stages of development. The refinements were a function of more material available and better instrumentation (embryos at the earliest stages

10. Haekel, *Anthropogenie*, 256.

are almost invisible to the naked eye). Had the *Science* article compared Richardson's photos with illustrations from Haeckel's later editions, the argument for fraud would have withered.

What about the considerable disparity between the images in the first edition of Haeckel's book and the photographs by Richardson and his colleagues? Even with the exculpating logical and historical considerations I have mentioned, how could a biologist of integrity represent a salamander embryo, looking like a lopsided beach ball in the photograph, as a slim, streamlined creature? It is that magnitude of difference that condemns Haeckel. But precisely here is the most dubious aspect of the case against him: several (but not all) of the photographed embryos retain the attached yolk sack and other maternal material; this exaggerates their differences from Haeckel's images. Haeckel explicitly indicated that he pictured his specimens without yolk, allantois, and amnion.[11] The bulge of the salamander in the photo is not part of the embryo; rather, it is the yolk sack, as is the case for the fish and the human embryos (though not for the chick and the rabbit, from which the yolk sacks have been removed). Moreover, the salamander photo is obviously not reduced to the same scale as the others (despite the assertion in the caption for the figure in *Science*). The chick was photographed in a highly circumflex orientation, which occurs at a somewhat later stage of development than that represented by Haeckel. Again, Haeckel expressly stated that he oriented his embryos all in the same way for ease of comparison. I have used a computer program to remove the yolks in the photographs, scale back the salamander, and straighten out the chick (fig. 7.5). The result is a bit crude, but one can clearly see that the differences between photograph and illustration are not nearly as great as presented in the *Science* article. Shorn of yolk, the photographed embryos would not have provided the kind of graphic evidence upon which the *Science* article was premised.

Haeckel was a man of great genius and driving passion. At times his impulsive energies led him astray, and he gave his opponents some cause for their complaints. In the first edition (1868) of his wildly popular *Natürliche Schöpfungsgeschichte* (Natural history of creation), he used the same woodcut three times to represent the initial formation of embryos of dog, chicken, and turtle. When a reviewer noticed this,[12] Haeckel defended himself by arguing that one could not tell the difference among these vertebrates at this very early

11. Ibid.

12. Ludwig Rütimeyer, "Review of *Natürliche Schöpfungsgeschichte* by Ernst Haeckel," *Archiv für Anthropologie* 3 (1868): 301–2.

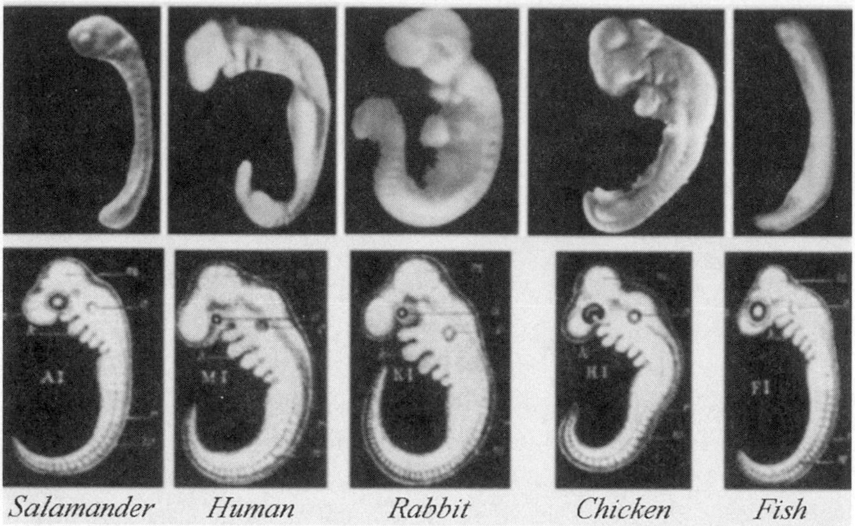


FIGURE 7.5 Reengineered photographs of embryos in fig. 7.1 with yolk material removed, comparable scaling, and reorientation.

stage—and given the instrumentation at the time, this was true. He nonetheless recognized that he egregiously erred and immediately corrected the text in the next edition two years later. The damage was done, however, and his enemies never ceased to remind readers of his misstep. Even with this stumble, however, he did not lose the support of such stalwarts as Darwin, Huxley, Weismann, and Gegenbaur. When Haeckel's science is placed in the wider context of his particular circumstances and the times, as I have attempted to do in my recent intellectual biography, his accomplishments appear in a decidedly more favorable light. And in the particular instance reviewed here, I think that light shows fraud has not been proven.[13]

13. I am grateful to Jerry Coyne who encouraged me to write this essay, based on my book Richards, *The Tragic Sense of Life* (Chicago: University of Chicago Press, 2008), and who patiently made many suggestions for improvement. I also owe thanks to my graduate students—Christopher DiTeresi, Alessandro Pajewski, and Trevor Pearce—who initially pointed out the discrepancies in the photographs.

CHAPTER EIGHT

# The Linguistic Creation of Man

## *August Schleicher and the Missing Link in Darwinian Theory*

While reflecting on various aspects of his new theory of species transformation, Charles Darwin (1809–1882) conjured up a singing ape and then one groaning its desires while eyeing a well-proportioned member of the opposite sex. Such utterances, he mused, may have been the phonetic resources for primitive speech. The problem of language had captured Darwin's attention from a quite early period in his theorizing about species descent. His initial concern was to show that language, that most human of traits, had a natural origin and that it developed in genealogical and progressive fashion. In a collection of notes, which he jotted down in 1837, shortly after returning from the *Beagle* voyage, he reflected on these putative features of language. On the very first page of this collection, he wrote: "all speculations on the origins of language—must presume it originates slowly—if these speculations are utterly valueless—then argument fails—if they have, then language was progressive.—We cannot doubt that language is an altering element, we see words invented—we see their origin in names of People—Sounds of words—argument of original formation.—declensions &c often show traces of origin."[1]

Language as a progressive achievement suggests a mundane rather than a divine origin. A bit later Darwin thought of that harmonious ape when he queried himself: "Did our language commence with singing"? Were we originally like howling monkeys or chirping frogs? But perhaps words arose out of

1. Charles Darwin, *Old and Useless Notes* (MS pp. 5 and 5v), in *Charles Darwin's Notebooks, 1836–1844*, ed. P. Barrett et al. (Ithaca: Cornell University Press, 1987), 599.

emotional responses to particular occasions (e.g., the ape with the opposite sex on its mind) or maybe from efforts at imitation of natural sounds.[2] These latter were the kinds of conjectures that Friedrich Max Müller (1823–1900; fig. 8.1), the great Oxford linguist, would later derisively call the "pooh-pooh" and "bow-wow" theories of language formation.[3] Darwin worried, even at this early juncture, that if his views about language origins could not be sustained, then his whole argument regarding evolution might fail, since that argument could not then explain one of the essential traits of human beings.

For the evolutionary thesis, no other trail lay open than the one Darwin initially began to follow. In the late 1860s, while focusing more determinately on constructing a theory of language, he came to rely in particular on his brother-in-law Hensleigh Wedgwood (1803–1891), who had endorsed a quasi-naturalistic account of linguistic development in his *On the Origin of Language* (1866). While working on what would become the *Descent of Man and Selection in Relation to Sex* (1871), Darwin made frequent inquiries of his cousin about this most curious subject. Wedgwood had allowed that it was part of God's plan to have man instructed, as it were, by the natural development of speech. He argued that language began from an instinct for imitation of the sounds of animals and natural events, which under "pressure of social wants" developed into a system of signs. According to Wedgwood, onomatopoeia served as the *vera causa* for a natural evolution of language.[4] Darwin embraced this confirmation of his original ideas, though dispensing with the theological interpretation. In the *Descent of Man*, he mustered this naturalistic account of language acquisition to a surprising purpose.

The principal concern of the *Descent of Man*, as the title signals, is the evolution of the human animal, with all of its distinctive properties, especially that

2. See Charles Darwin, *Notebook N* (MS pp. 18, 65), in Barrett et al., *Charles Darwin's Notebooks*, 568, 581.

3. In his lectures on language, Max Müller dismissed the "the Bow-wow theory and the Pooh-pooh theory" of language origins. He argued that the syllabic roots of language, the elements of the various language families, "are not interjection, nor are they imitations. They are *phonetic types* produced by a power inherent in human nature. They exist . . . by nature; . . . when we say by nature, we mean by the hand of God." Once these roots had been planted and expressed in an instinctual fashion, human reason would gradually begin to use them for ideas. Rather cleverly, Max Müller borrowed Darwin's device of natural selection to argue that the shaping of language in social settings occurred through selection against certain forms. See Max Müller, *Lectures on the Science of Language* (London: Longman, Green, Longman, and Roberts, 1861), 344–56, 370–71.

4. See Hensleigh Wedgwood, *On the Origin of Language* (London: Trübner, 1866), 13–14, 129. See also Robert J. Richards, *Darwin and the Emergence of Evolutionary Theories of Mind and Behavior* (Chicago: University of Chicago Press, 1987), 205.

FIGURE 8.1 Friedrich Max Müller (1823–1919), in 1857.
Photo by Charles Dodgson (Lewis Carroll). (© National Portrait Gallery)

of high intellect. Yet, Darwin agreed with his friend Alfred Russel Wallace (1823–1813; fig. 8.2) that for survival alone, man's apelike ancestors needed a brain hardly larger than that of an orangutan. Wallace was reinforced in this conclusion by his unexpected turn toward spiritualism. He came to believe that man's ascent from the animal state occurred through the ministrations of

FIGURE 8.2 Alfred Russel Wallace (1823–1913), ca. 1864. Oil over photography by Thomas Sims. (© National Portrait Gallery)

higher, spiritual powers—a proposal that drove Darwin crazy.[5] Darwin nonetheless recognized the force of Wallace's objection. If a large brain were not required for survival, then natural selection could not account for it. Darwin thus needed another way to explain the refinement and perfection of human intelligence. Language provided the instrument, though not in the way we might acknowledge today. In the *Descent of Man*, he argued in this fashion:

> The mental powers in some early progenitor of man must have been more highly developed than in any existing ape, before even the most imperfect form of speech could have come into use; but we may confidently believe that the continued use and advancement of this power would have reacted on the mind by enabling and encouraging it to carry on long trains of thought. A long and complex train of thought can no more be carried on without the aid of words, whether spoken or silent, than a long calculation without the use of figures or algebra.[6]

Darwin proposed that our apelike ancestors must have developed considerable intellectual capacity prior to breaking into the human range of intelligence. That animals displayed conspicuous understanding, approaching that of the human, no English huntsman seriously doubted. Even the great British Idealist F. H. Bradley (1846–1924) remarked to a friend: "I never could see any difference at bottom between my dogs & me, though some of our ways were certainly a little different."[7] (This may say more about late nineteenth-century British philosophy than about the abilities of English canines.) What was needed, in Darwin's view, to steam our animal ancestors across the Rubicon of mind was the engine of language. As language evolved through a natural development out of emotional and imitative cries, it would rebound on brain, promoting, as Darwin indicated, a more complex train of thought. Darwin would differ from contemporary neo-Darwinians, however. He believed that the complex patterns of thought that language stimulated would progressively alter brain structures and that these new acquisitions would produce an

5. I have discussed Wallace's spiritualistic interpretation of evolution and Darwin's reaction in Richards, *Darwin and the Emergence of Evolutionary Theories of Mind and Behavior*, 176–84. See also chapter 4 of the present volume. Wallace made a spirited defense of the reality of mediumship and other psychic phenomena in Wallace, "A Defense of Modern Spiritualism," *Fortnightly Review*, n.s., 15 (1874): 630–57, 785–807.

6. Charles Darwin, *On the Descent of Man and Selection in Relation to Sex*, 2 vols. (London: Murray, 1871), 1:57.

7. F. H. Bradley to C. Lloyd Morgan (16 February 1895), in the Papers of C. Lloyd Morgan, DM 612, Bristol University Library.

"inherited effect."[8] Darwin thus contended that language created human brain and, consequently, human mind.[9]

From the beginning of his career to the end, Darwin believed in the inheritance of acquired characteristics. From our current perspective, we can see that he need not have argued in this fashion. He could have employed his own device of natural selection to explain the reciprocal pressures that mind and language might have exerted on one another to produce a continued coevolution of both.[10] Darwin did not appreciate that ever more complex language and thought might have had distinct survival advantages—for example, even rudimentary language might have served to weave together mutually supportive social networks of our protohuman ancestors. Like Wallace, he conceded that for sheer survival our progenitors did not require a brain more advanced than that of, say, a great ape. Hence, in those cases in which natural selection seemed inapplicable, Darwin fell back on that device he always had at the ready—the inheritance of acquired characters.

Darwin's theory of the influence of language on developing mentality seems, at first blush, puzzling. This is not because of his employment of the idea of use-inheritance—common enough for his theory and his time. The puzzle rather arises because his proposal ran counter to the usual British empiricists' assumption that language merely expressed or mirrored ideas—it did not create them.[11] What then was the source of Darwin's conviction that language could mold human brain, could create human mind? In what follows I argue that the ultimate source for his conception is to be found in German Romanticism and Idealism, especially in the work of Wilhelm von Humboldt (1767–1835), linguist and pedagogical architect of the University of Berlin, and of Georg Friedrich Hegel (1770–1831), Germany's greatest philosopher at midcentury. German Romanticism and Idealism thus forged, I believe, a missing link in nineteenth-century evolutionary theory.

8. Darwin, *Descent of Man*, 1:58, 2:390–91.

9. Once brain became further refined through language use, other forces would aid in its construction, namely, community selection. See chapter 4 in this volume.

10. For a wide-ranging and compelling discussion of the language–brain relationship, see Terrence Deacon, *The Symbolic Species: The Co-Evolution of Language and the Brain* (New York: Norton, 1997).

11. John Locke (1632–1704), as usual, established the common British view. He held that God furnished man with language in order "to use these sounds as signs of internal conceptions; and to make them stand as marks for the ideas within his own mind, whereby they might be made known to others, and the thoughts of men's minds be conveyed from one to another." Though thought used language, according to Locke "thought is not constituted by, nor identical with language, which on the contrary is originated and formed by thought." See John Locke, *An Essay concerning Human Understanding* (1670), 2 vols. (New York: Dover, 1959), 2:3 and n. 2.

## DARWIN AND THE LINGUISTIC RUBICON

Although Darwin realized that he would have to give an account of human mind and language if his general theory were to win the day, he kept all overt discussion of human evolution out of the book that first detailed his theory, the *Origin of Species* (1859). He simply forecast in the concluding chapter that "light will be thrown on the origin of man and his history."[12] The *Origin* is, nonetheless, larded with oblique but succulent references to human activity and history.[13] The case of language stands out among these. In his chapter on classification and systematics, for instance, Darwin observed: "If we possessed a perfect pedigree of mankind, a genealogical arrangement of the races of man would afford the best classification of the various languages now spoken throughout the world; and if all extinct languages, and all intermediate and slowly changing dialects, had to be included, such an arrangement would, I think, be the only possible one."[14]

In this passage, Darwin recognized an isomorphism between language descent and human biological descent. So not only could the human pedigree serve as a model for tracing linguistic development, as he here emphasized, but the reverse, as he also implied, could be the case as well: the descent of language might serve as a model for the descent of man.

Darwin's suggestion about a similar genealogy for human beings and language passed casually through only one paragraph of the *Origin*.[15] He himself did not really employ the model in any systematic way, and the paragraph seems almost an afterthought. The bare suggestion of this apparent isomorphism between the development of language and the development of human varieties, however, caught fire almost immediately. For the moment, Darwin warmed himself contentedly in the glow, until, that is, Lyell threw in what initially seemed supportive considerations but which ultimately proved quite threatening.

Charles Lyell (1797–1875; fig. 8.3) was Darwin's longtime friend and a scientist out of whose brain, Darwin said, came half his own ideas. Lyell immediately took up Darwin's suggestion about descent of language and further advanced it in his book *The Antiquity of Man* (1863). Lyell had observed that though there were wide gaps between dead and living languages, with no transitional

12. Charles Darwin, *On the Origin of Species* (London: Murray, 1859), 488.

13. See Kathy J. Cooke, "Darwin on Man in the *Origin of Species*," *Journal of the History of Biology* 26 (1990): 517–21.

14. Darwin, *Origin of Species*, 422.

15. Ibid., 422–23.

FIGURE 8.3 Sir Charles Lyell (1797–1875), in 1855. Lithograph of portrait, 1849; from the author's collection.

dialects preserved, competent linguists did not doubt the descent of modern languages from ancient ones. Therefore, gaps in the fossil record of species ought to prove no more of an obstacle to transmutation theory than gaps in the record of languages proved in linguistic theory. Moreover, he believed that the two kinds of descent should have a common explanatory account. So the formation and proliferation of languages were due, to quote Lyell, to "fixed laws in action, by which, in the general struggle for existence, some terms and dialects gain the victory over others."[16] Lyell thus maintained that the processes

16. Charles Lyell, *The Geological Evidences of the Antiquity of Man* (London: Murray, 1863), 463.

of biological evolution could be likened to those of linguistic evolution—in both the more fit types were selected. Lyell, one of Britain's leading scientists of the time, thus offered significant support for his friend's theory.

Lyell, however, could not make it across the Rubicon. He thought the principle of natural selection unable to account completely for the intricately designed fabric of language, even that of the more primitive languages of native groups. He judged—as Darwin groaned his great frustration—that natural selection of both language and life-forms could only be a secondary cause, operating under the guidance of more general laws. "If we confound 'Variation' or 'Natural Selection' with such creational laws," he cautioned, "we deify secondary causes or immeasurably exaggerate their influence."[17] Lyell repaired, quite obviously, not to natural law, but to a Divine Will. The turn to an interventionist God eviscerated Darwinian nature of the fecund force with which the *Origin* invested it. Nature, in Darwin's theory, resonated with that romantic power of creative action and evaluation that it soaked up from German sources, especially from Alexander von Humboldt (1769–1859), whom Darwin incessantly read while on the *Beagle* voyage some years before.[18] But another German writer came to Darwin's attention in the mid-1860s, one whose analyses of language he found considerably more congenial than Lyell's and whose ideas he would weave into his own theory of human evolution. This was August Schleicher (1821–1868).

## AUGUST SCHLEICHER'S LINGUISTIC DARWINISM

Schleicher (fig. 8.4) was a distinguished linguist working at the university in Jena. He had been urged by his good friend Ernst Haeckel (1839–1919) to read the German edition of the *Origin*. Haeckel, who himself had recently converted to Darwinism, recommended the book because of Schleicher's

17. Ibid., 469.

18. Alexander von Humboldt not only conveyed a conception of living nature that Darwin incorporated into his own evolutionary theory, but he also suggested that language helped to create human intellect. In the English translation of Humboldt's *Kosmos*, which Darwin read in the 1850s, the following may be found: "But thought and language have ever been most intimately allied. If language, by its originality of structure and its native richness, can, in its delineations, interpret though with grace and clearness, and if, by its happy flexibility, it can paint with vivid truthfulness the objects of the external world, it reacts at the same time upon thought, and animates it, as it were, with the breath of life. It is this mutual reaction which makes words more than mere signs and forms of thought; and the beneficent influence of a language is more strikingly manifested on its native soil, where it has sprung spontaneously from the minds of the people, whose character it embodies." See Alexander von Humboldt, *Cosmos*, 5 vols., trans. E. C. Otté (New York: Harper and Brothers, 1848–68), 1:56. This general view of language is also found in August Schleicher, as I explain later in the chapter. Both theorists, however, seem to have had a common source: Wilhelm von Humboldt.

FIGURE 8.4 August Schleicher (1821–1868). Engraving. (Berlin/Preussischer Kulturbesitz/Art Resource, NY)

horticultural interests. The linguist was a serious gardener and quickly wrote a review of the book for an agricultural journal.[19] In the review Schleicher summarized Darwin's argument and added his own extension of the theory, proposing that human beings descended from the "higher apes," differing from the apes principally in language and larger brain capacity. He didn't mention that Darwin in the *Origin* hadn't explicitly discussed the evolution of man. But already Schleicher was undoubtedly thinking about how language mattered in the transition from animals to man.

Virtually at the same time as his review, Schleicher set quickly to respond to Darwin's work in the manner of the linguist. He wrote an open letter to his colleague Haeckel in the form of a small tract: *Die Darwinsche Theorie und die Sprachwissenschaft* (Darwinian theory and the science of language, 1863).[20] The little book excited considerable controversy, evoking critically negative responses from the likes of Friedrich Max Müller and the American linguist William Dwight Whitney (1827–1894), but supportive efforts from the

19. See August Schleicher, "Die Darwin'sche Theorie und die Thier- und Pflanzenzucht," *Zeitschrift für deutsche Landwirthe* 15 (1864): 1–11. Schleicher sent the review to Darwin, and it is now held in the Department of Manuscripts, Cambridge University Library. Scorings indicated Darwin read the review.

20. August Schleicher, *Die Darwinsche Theorie und die Sprachwissenschaft* (Weimar: Böhlau, 1863). See also two works that discuss Schleicher's little book: Liba Taub, "Evolutionary Ideas and 'Empirical' Methods: The Analogy between Language and Species in Works by Lyell and Schleicher," *British Journal for the History of Science* 26 (1993): 171–93; and Stephen Alter, *Darwinism and the Linguistic Image* (Baltimore: Johns Hopkins University Press, 1999), esp. 73–79.

British scholar Frederick Farrar (1831–1903).[21] In the *Descent of Man*, Darwin referred to his brother-in-law Hensleigh Wedgwood and Farrar as sources for his ideas about the evolutionary descent of language. He silently prescinded, as one might expect, from the fact that each of his sources reserved a role for the Creator. And he credited Schleicher as well. It was on Schleicher's thoroughgoing linguistic naturalism, I believe, that Darwin principally depended for his theory of the constructive effect of language on mind.

Schleicher indicated that contemporary languages had gone through a process in which simpler *Ursprachen* had given rise to descendant languages that obeyed natural laws of development. He argued that Darwin's theory was thus perfectly applicable to languages and, indeed, that evolutionary theory itself was confirmed by the facts of language descent. This last point was crucial for Schleicher, since it suggested the singular contribution that the science of language could make to the establishment of Darwin's theory. In the German translation of the *Origin*, Heinrich Georg Bronn (1800–1862), the translator, had added an epilogue in which he allowed that Darwin's theory showed that descent was possible but that the Englishman had not shown it was actual. Darwin had, according to Bronn, no direct empirical evidence, only analogical possibilities.[22] Schleicher, like many other Germans, accepted Bronn's evaluation.

21. In the English press, Schleicher's book, in the first German edition, received immediate notice through an anonymous author. See "The Darwinian Theory in Philology," *Reader* 3 (1864): 261–62. The author agreed with Schleicher that linguistics lent support to Darwin's theory. Friedrich Max Müller discussed the English translation of the work in a review in *Nature*: "The Science of Language," *Nature* 1 (1870): 256–59. Müller took exception to the idea that descendant languages sprang from a well-formed classical language (e.g., French from Latin). He rather maintained that the descendant languages arose from rude dialects that might trace their origin to the classical language. Frederick Farrar, believing that Müller gave scant account of Schleicher's little book, provided a summary in a subsequent issue of *Nature*: "Philology and Darwinism," *Nature* 1 (1870): 527–29. Darwin undoubtedly read these reviews. See the discussion of the controversy between Müller and Farrar in Alter, *Darwinism and the Linguistic Image*, 84–96. William Dwight Whitney took grave exception to Schleicher's naturalism—that is, the supposition that languages displayed organic features and obeyed natural laws—and denied that Schleicher's notion of language descent gave any aid to Darwin's theory. See William Dwight Whitney, "Schleicher and the Physical Theory of Language" (1871), reprinted in his *Oriental and Linguistic Studies*, 2 vols. (New York: Charles Scribner's Sons, 1873), 1:298–331. Hans Arsleff details other responses to Schleicher's *Darwinsche Theorie*, of whose doctrines he himself thoroughly disapproves. See Hans Arsleff, *From Locke to Saussure: Essays on the Study of Language and Intellectual History* (Minneapolis: University of Minnesota Press, 1982), 293–334.

22. Heinrich Bronn, "Schlusswort des Übersetzers," in Charles Darwin, *Über die Entstehung der Arten im Thier- und Pflanzen-Reich durch natürliche Züchtung oder Erhaltung der vervollkommenten Rassen im Kampfe um's Daseyn* (from the 2ne English ed.), trans. H. G. Bronn (Stuttgart: Schweizerbart'sche Verlagshandlung und Druckerei, 1863), 493–520. Bronn brought as a chief objection to Darwin's theory that it was "in its ground-conditions of justification still a thoroughly wanting hypothesis." It remained, according to Bronn, "undemonstrated," though also "unrefuted" (502). Bronn did, however, lodge some considerations that militated against the hypothesis, for example, that transitional species were lacking (504). Bronn himself was the author of a quasi-evolutionary theory, which he formulated prior to reading Darwin. He elaborated

He yet insisted that language descent, unlike the imaginative scenarios Darwin offered, could be proved—it was already an empirically established phenomenon. Moreover, the linguist's descent trees (*Stammbäume*) might be used as models for construing the evolution of plant and animal species.

Schleicher was quick to point out that the only graphic representation of descent in Darwin's *Origin* consisted of a highly abstract scheme in which no real species were mentioned, only letter substitutes (fig. 8.5). He contrasted this with a descent tree of the Indo-Germanic languages—his own graphic innovation—which he attached as an appendix to his tract (fig. 8.6). Darwin had thus only represented a possible pattern of descent, while the linguist could provide a real pattern, empirically derived. Here, Schleicher believed, was a genuine contribution of linguistics to biological theory, a contribution that undercut Bronn's objection.

Schleicher maintained there were some four other areas in which the linguistic model could advance the Darwinian proposal. First, the linguistic system might display a "natural history of the genus homo." This is because "the developmental history of languages is a main feature of the development of human beings." Second, "languages are natural organisms [Naturorganismen]" but have the advantage over other natural organisms since the evidence for earlier forms of language and transitional forms has survived in written records—there are considerably more linguistic fossils than geological fossils. Third, the same processes of competition among languages, the extinction of forms, and the development of more complex languages out of simpler roots all suggest mutual confirmation of the basic processes governing such historical entities as species and languages. Finally, since the various language groups descended from "cellular languages," language provides analogous evidence that more advanced species descended from simpler forms.[23]

---

his theory in a prizewinning essay, selections of which were translated into English as "On the Laws of Evolution of the Organic World during the Formation of the Crust of the Earth," *Annals and Magazine of Natural History*, 3rd ser., 4 (1859): 81–90, 175–84. Bronn argued for a gradual appearance of new species and an extinguishing of more primitive ones over great periods of time. Such evolution , however, did not involve the transformation of one species into another but merely the successive appearance and adaptation of progressively higher kinds of flora and fauna. This process occurred, he strongly implied but did not expressly say, through Divine Wisdom. His views were not unlike those of Louis Agassiz and Richard Owen. For a discussion of naturalists like Bronn who had a developmental theory of descent (but not a genealogical one) prior to Darwin, see Nicolaas Rupke, "Neither Creation nor Evolution: The Third Way in Mid-Nineteenth Century Thinking about the Origin of Species," *Annals of the History and Philosophy of Biology* 10 (2005): 143–72.

23. Schleicher, *Darwinsche Theorie*, 4–8, 23–24.

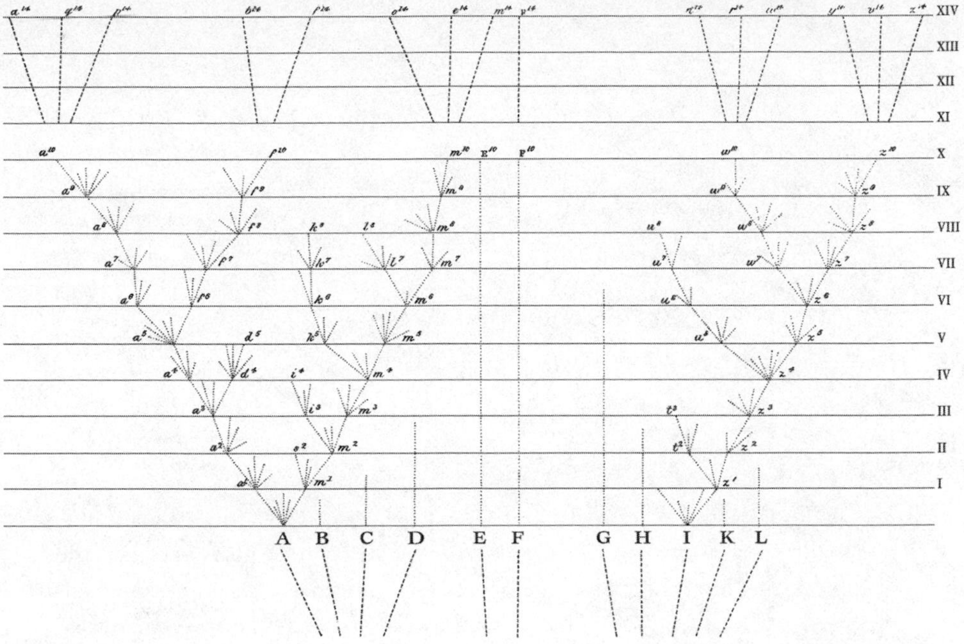


FIGURE 8.5 DARWIN'S schema for species descent in the *Origin of Species.*

Schleicher intended that these four complementary contributions of linguistics to biological theory should buttress an underlying conviction of his *Darwinsche Theorie*, namely, that the pattern of language descent perfectly reflected that of human descent. The implicit justification for this proposition was simply that these two processes of descent were virtually the same—an idea I will explore further in a moment. And this justification itself was grounded in the doctrine of monism that Schleicher advanced in his tract. That doctrine, as he formulated it, recognized that

> Thought in the contemporary period runs unmistakably in the direction of monism. The dualism, which one conceives as the opposition of mind and nature, content and form, being and appearance, or however one wishes to indicate it—this dualism is for the natural scientific perspective of our day a completely unacceptable position. For the natural scientific perspective there is no matter without mind [Geist] (that is, without

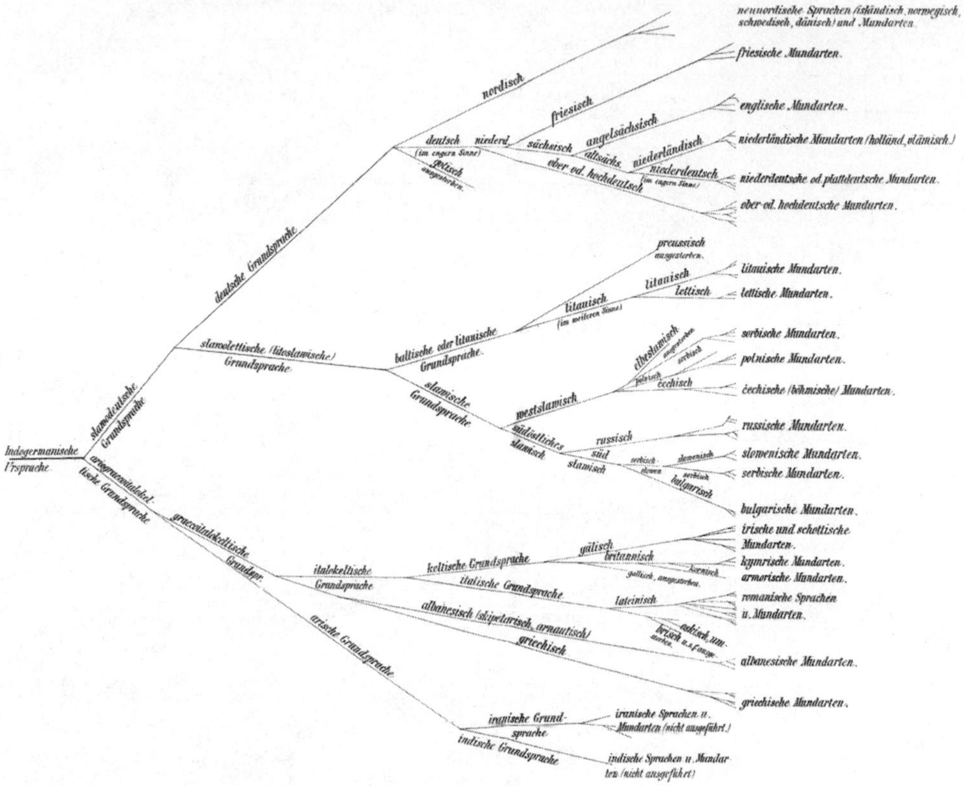


FIGURE 8.6 August Schleicher's diagram of the descent relations of the Indo-Germanic languages. From his *Darwinische Theorie und die Sprachwissenschaft* (1863).

that necessary power determining matter), nor any mind without matter. Rather there is neither mind nor matter in the usual sense. There is only one thing that is both simultaneously.[24]

For Schleicher, the doctrine of monism provided a metaphysical ground for his theory that the organism of language simply represented the material side of mind—which meant, therefore, that the evolution of one carried the evolution of the other. This organic naturalism had its roots in the German Romantic movement. That movement rejected the mechanistic interpretation

24. Ibid., 8.

of nature and advanced the concept of *organism* as the fundamental principle in terms of which human mentality and all natural phenomena were ultimately to be understood.

In a small work published two years after *Darwinsche Theorie*, Schleicher developed some further features of his complementary theories of linguistic and human evolution. In *Über die Bedeutung der Sprache für die Naturgeschichte des Menschen* (On the significance of language for the natural history of mankind, 1865), he argued that the superficial differences among human beings, which morphologists often exaggerated, proved simply insufficient for classification. He observed:

> How inconstant are the formation of the skull and other so-called racial differences. Language, by contrast, is always a constant trait. A German can indeed display hair and prognathous jaw to match those of the most distinctive Negro head, but he will never speak a Negro language with native facility. . . . Animals can be ordered according to their morphological character. For man, however, the external form has, to a certain extent, been superseded; as an indicator of his true being, external form is more or less insignificant. To classify human beings we require, I believe, a higher criterion, one which is an exclusive property of man. This we find, as I have mentioned, in language.[25]

Schleicher contended that some languages were more developed than others and that this fact could provide a progressive arrangement of the human races. He believed, perhaps not surprisingly, that the Indo-Germanic and Semitic language groups were the most advanced, since they had features, such as tenses, declensions, and true noun and verb forms, which were lacking in such languages as Chinese. By implication, he thus suggested that the most evolved human groups in the evolutionary hierarchy were those whose native languages were of the Indo-Germanic and Semitic families. Schleicher's justification for using language to classify human groups was simple: "the formation of language is for us comparable to the evolution of the brain and the organs of speech."[26] This was the position Darwin endorsed, and it became for him a central feature of his evolutionary conception of mankind (discussed later in the chapter).

25. August Schleicher, *Über die Bedeutung der Sprache für die Naturgeschichte des Menschen* (Weimar: Böhlau, 1865), 16, 18–19.

26. Ibid., 21.

Schleicher claimed that he himself had been convinced of the natural descent and competition of languages before he had read the *Origin of Species*. While it is difficult to corroborate his assertion that he had previously urged a *Kampf ums Dasein* (struggle for existence) to explain language change, there is little doubt that he had affirmed language competition and descent as natural phenomena prior to reading Darwin and that he subsequently used these concepts to argue for human evolution. Schleicher's argument, however, displays fascinating archeological layers of earlier ideas.

## ORIGIN OF SCHLEICHER'S EVOLUTIONARY THEORY OF LANGUAGE AND MIND

Schleicher was born 19 February 1821 in Meiningen (southwest of Weimar in the Thuringian Forest) to a physician with a taste for nature and his musically talented wife.[27] The professors of his gymnasium cultivated exotic languages but did not, amazingly, have high hopes for this particular pupil. In fall 1840, Schleicher began the curriculum in theology at Leipzig and the next semester traveled to Tübingen for more of the same. At Tübingen his passion for the transcendent found secular liberation in Hegel's writings, which had been recently collected by his students (1832–40), with many works appearing for the first time. Schleicher also began acquiring languages at a frightening rate: Arabic, Hebrew, Sanskrit, and Persian initially. With the reluctant permission of his father, he went to Bonn in 1843 to devote himself to the study of classical languages. There he entered the seminar conducted by the famous classical philologists Friedrich Ritschl (1806–1876) and Friedrich Welcker (1784–1868), who introduced him to the linguistic ideas of Wilhelm von Humboldt.[28] Though of oscillating health while at Bonn, Schleicher yet braced his study with participation in gymnastic competitions, a recreation that he and Haeckel would later pursue together with avidity. He received a doctorate in 1846 and would normally have then spent time as a professor in a gymnasium before pursing further study. He fell, however, under the protective wing of Prince Georg von Meiningen, who, admiring of his landsman's talents, arranged for a generous stipend. The money enabled Schleicher to continue his study during a period of two years of extensive travel (1848–50).

27. For details of Schleicher's life I have relied on Johannes Schmidt, "Schleicher," *Allgemeine deutsche Biographie* 31 (1890): 402–15; Joachim Dietze, *August Schleicher als Slawist: Sein Leben und sein Werk in der Sicht der Indogermanistik* (Berlin: Adademie-Verlag, 1966); and Theodor Syllaba, *August Schleicher und Böhmen* (Prague: Karls-Universität, 1995).

28. Schleicher, *Bedeutung der Sprache*, 18.

In the summer of 1848, after the February Revolution and the establishment of the Second Republic, Schleicher journeyed to Paris to continue his linguistic research at the Bibliothéque Nationale. He augmented his income during this sojourn by serving as correspondent to *Allgemeine Zeitung* (Augsburg) and *Kölnische Zeitung*. He reported on the fluctuating political events occurring in Paris and a bit later in Vienna, as revolution spread to the capital of the Hapsburg Empire. Schleicher's reports, tinged with the sympathetic color of a liberal democrat, followed the fate and abortive efforts to establish a republic in the Germanies.[29] In addition to his political reporting, Schleicher managed to produce a number of important linguistic studies, which elicited a call from the University of Prague to the position of extraordinary professor. Three years later, he advanced to ordinary professor of German, comparative linguistics, and Sanskrit. He remained in Prague until 1857, when he received an offer to return to his own land. He accepted a position in the philosophy faculty at Jena, the venerable university that two generations' earlier, at the turn of the century, had nurtured the Romantic movement, serving as redoubt for the likes of Schiller, Fichte, the brothers Schlegel, Schelling, Hegel, and with Goethe right down the road at Weimar.[30] Jena was also the university of Schleicher's father, Johann Gottlieb Schleicher (1793–1864), who in the summer of 1815 helped found the first *Burschenschaft*, the student organization that agitated for democratic reform and political unity.[31] In the 1850s, the university looked back to a glorious past and forward to a financially precarious future.

Although he initially had high hopes for his time in Jena, undoubtedly recalling his father's stories of revolutionary days at the university, Schleicher quickly came to feel isolated from his colleagues, whose conservative considerations bent them away from the more daring of his own approaches both in linguistics and politics. The poor finances of the university, which made scarce the necessities for scholarship, did not improve his attitude. A friend remembered Schleicher remarking that "Jena is a great swamp and I'm a frog in it."[32] The frog was saved from wallowing alone in his pond when Ernst Haeckel arrived at the university in 1861. They took to one another immediately and remained fast friends through the rest of Schleicher's short life. He died in 1868, at age forty-eight, apparently of a recurrence of tuberculosis.

29. Syllaba characterizes Schleicher's work as a correspondent and provides a list of the articles, in Syllaba, *August Schleicher und Böhmen*, 13–27.

30. I have detailed the history of the early Romantic movement in Germany and the roles of the aforementioned figures in Robert J. Richards, *The Romantic Conception of Life: Science and Philosophy in the Age of Goethe* (Chicago: University of Chicago Press, 2002).

31. See Dietze, *August Schleicher als Slawist*, 16.

32. Robert Boxberger, "Prager Erinnerungen aus Jena," as quoted ibid, 45.

In 1848—after he returned to Bonn from research in the revolution-torn city of Paris—Schleicher saw published his first monograph, *Zur vergleichenden Sprachengeschichte* (Toward a comparative history of languages).[33] This work framed the theory that would guide him through the rest of his career. In it, he distinguished three large language families by reason of their forms: isolating languages, agglutinating languages, and flexional languages. Isolating languages (e.g., Chinese and African) have very simple forms, in which grammatical relationships are not expressed in the word; rather, the word consists merely of the one-syllable root (with position or pitch indicating grammatical function). Because of their simple structure, these languages cannot, according to Schleicher, give full expression to the possibilities of thought. Agglutinating languages (e.g., Turkish, Finnish, Magyar) have their relational elements tacked on to the root in a loose fashion (indeed, the relational elements themselves are derived from roots). Flexional languages (e.g., the Indo-Germanic and Semitic families) are the most developed. Roots and grammatical relations form an "organic unity," according to Schleicher.[34] So, for example, the Latin word *scriptus* has *scrib* as the root or meaning; *tu* expresses the participial relationship; and *s* indicates the nominative relationship. Schleicher believed that even the most highly developed languages, the flexional group, originated from a simpler stem, much like the Chinese, but continued to develop into varieties with more perfect forms. Isolating and agglutinating languages, on the other hand, simply did not have the potential to move much beyond their more primitive structures.

Schleicher regarded these three language forms as exhibiting an internal, organic unity. Indeed, he compared them to natural organisms of increasing complexity: crystals, plants, and animals, respectively.[35] Such comparisons had the authority of those linguists upon whom Schleicher most relied: Wilhelm von Humboldt, Franz Bopp (1791–1867), and August Wilhelm Schlegel (1767–

33. August Schleicher, *Zur vergleichenden Sprachengeschichte* (Bonn: H. B. König, 1848).

34. In distinguishing these three forms of language, Schleicher was simply following the lead of Wilhelm von Humboldt, Franz Bopp, and ultimately August Wilhelm Schlegel. Schleicher was certainly familiar with the work of these near contemporary linguists. In his *Sprachengeschichte*, Schleicher cited Humboldt often enough, though not precisely on this distinction. See Wilhelm von Humboldt, *Über die Kawi-Sprache auf der Insel Java*, 3 vols. (Berlin: Königlichen Akademie der Wissenschaften, 1836). The introduction to this famous work on Javanese language made the threefold distinction pivotal (1:cxxxv–cxlviii). August Wilhelm Schlegel, who became professor of linguistics at Bonn, formulated the original distinction in his *Observations sur la langue et la littérature provençales* (Paris: Librairie Grecque-Latine-Allemande, 1818), 14–16. Franz Bopp, whom Humboldt brought to Berlin as professor, canonized the distinction in his *Vergleichende Grammatik des Sanskrit, Zend, Griechischen, Lateinischen, Litthauischen, Gothischen und Deutschen* (Berlin: Königlichen Akademie der Wissenschaften, 1833), 108–13.

35. Schleicher, *Zur vergleichenden Sprachengeschichte*, 8–11.

1845). These researchers, all tinged by the Romantic movement, employed the organic metaphor with alacrity.[36] Schleicher, though, suggested an important disanalogy between languages and biological organisms. Languages had a developmental history, whereas biological organisms, though they came to exist through a gradual process, once established did not alter. They essentially had no history. At least this was Schleicher's view in 1848.

In 1850, Schleicher completed a large monograph systematically describing the languages of Europe, his *Die Sprachen Europas in systematischer Übersicht* (The languages of Europe in systematic perspective). He now explicitly represented languages as perfectly natural organisms that could most conveniently be described using terms drawn from biology—for example, genus, species, and variety.[37] Some of his contemporaries, as well as later linguists, thought Schleicher's conception of language as a natural, law-governed phenomenon to be erroneous, a denial of man's special status. Such critics then (and now) failed to understand that this was not a denigration of the *geistlich* character of language; rather, it was, in the Romantic purview, an elevation of the natural.[38] Romantics and Idealists—such as Schelling, Schlegel, and Hegel—deemed nature simply the projection of mind. Schleicher, then, did not reduce in vulgar fashion the spiritual dimension of language to some nonanimate concourse of atoms in the void.

36. Humboldt, for instance, liked to refer to the internal coherence of language by use of the term "the language-organism" (*Sprachorganismus*). See Humboldt, *Kawi-Sprache*, 1:cxxxv. Bopp likewise generously employed the organic metaphor; as he expressed it in his *Vergleichende Grammatik*: "I intend in this book a comparative, comprehensive description of the organism of the languages mentioned in the title, an investigation of their physical and mechanical laws, and the origin of the forms indicating grammatical relationships" (iii). Humboldt and Bopp, in utilizing this metaphor, adopted the conception of Friedrich Schelling, the philosophical architect of the Romantic movement and one who made the organic a controlling principle of mind and matter, roughly from 1798 on. See, for instance, a typical observation in his *Historisch-kritische Einleitung in die Philosophie der Mythologie* (1842), in Friedrich Wilhelm Joseph von Schelling, *Ausgewählte Schriften*, ed. Manfred Frank, 6 vols. (Frankfurt am Main: Suhrkamp, 1985), 5:61: "Language does not arise piece-meal or atomistically, but it arises in all its parts immediately as a whole and thus organically [organisch]."

37. Although Schleicher basically advanced the same theory as that in his *Sprachengeschichte*, he now felt perfectly comfortable describing language groups using biological classifications. See August Schleicher, *Die Sprachen Europas in systematischer Übersicht* (Bonn: König, 1850), 22–25, 30.

38. Among his contemporaries, William Dwight Whitney dismissed Schleicher's conception of language as a law-governed, organic phenomenon. Whitney argued that actions produced by human will escaped the rule of law. See Whitney, "Schleicher and the Physical Theory of Language," 298–331. This same kind of criticism has been voiced more recently. Eugen Seidel thinks Schleicher "erred" in regarding *Sprachwissenschaft* (linguistics) as a *Naturwissenschaft* (natural science), failing, as he supposedly did, to perceive the social character of language. See Eugen Seidel, "Die Persönalichkeit Schleichers," *Wissenschaftliche Beiträge der Friedrich-Schiller-Universität Jena* (1972): 8–17. Arsleff expresses a similar opinion (*From Locke to Saussure*, 294–95). Such judgments betray a poverty of historical understanding.

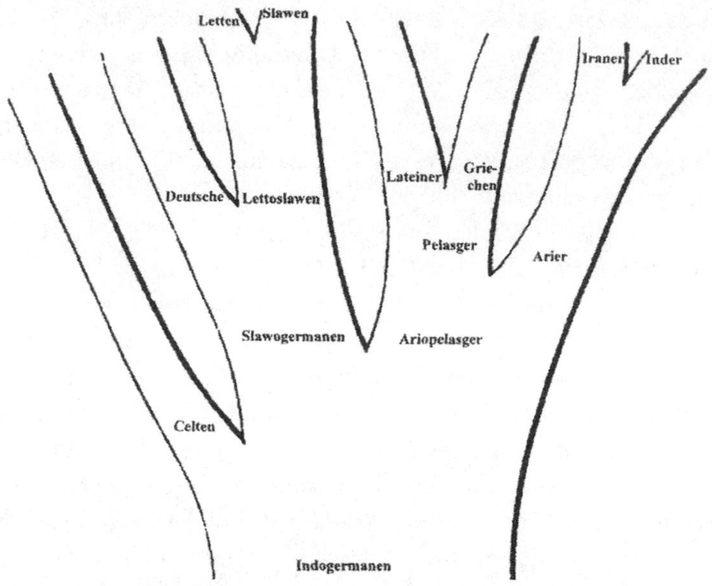


FIGURE 8.7 Schleicher's first diagram of language descent. From his "Die ersten Spaltungen des Indogermanischen Urvolkes" (1853).

In his *Sprachen Europas*, Schleicher suggested (but did not yet graphically illustrate) that the developmental history of the European languages could best be portrayed in a *Stammbaum*, a stem-tree or developmental tree. He first introduced a graphic representation of a *Stammbaum* in articles published in 1853; the representations indeed looked like trees (fig. 8.7).[39] By the time of the publication of his *Deutsche Sprache*, seven years later (1860), he had begun to use *Stammbäume* rather frequently to illustrate language descent (fig. 8.8).

Schleicher is commonly recognized as the first linguist to portray language development by using the figure of a tree.[40] Certainly he thought carefully about how illustrations could make more clear and more intuitive the descent relations that purportedly obtained among languages. So, for instance, he used the angular distance separating the branching of the *Stammbaum* to suggest the

39. Schleicher published two articles in 1853 that employed a graphic illustration of a *Stammbaum*. One was in Czech, the other German. See, for instance, August Schleicher, "Die ersten Spaltungen des indogermanischen Urvolkes," *Allgemeine Zeitschrift für Wissenschaft und Literatur* (August 1853): 786–87.

40. Taub thinks that Friedrich Ritschl (1806–1876), Schleicher's teacher at Bonn, may have suggested the tree method of representation by his work in the establishment of manuscript pedigrees. See Taub, "Evolutionary Ideas and 'Empirical' Methods," 185–86.

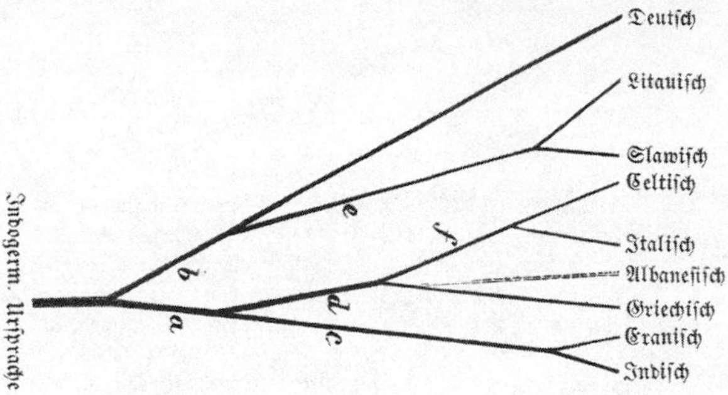


FIGURE 8.8 Schleicher's stem-tree of the Indo-Germanic languages. From his *Die Deutsche Sprache* (1860).

morphological distances of daughter languages (fig. 8.9).[41] Such illustrations, so intuitively seductive, acted as tacit arguments for the theory they depicted.

In *Deutsche Sprache*, Schleicher reiterated the argument of *Sprachen Europas* that more recent languages had descended from *Ursprachen* and that their descent conformed to natural laws. He now, however, started to formulate those laws; for example, "When two or more branches of a language stem [Sprachstamm] are quite similar, we may naturally conclude that they have not been separated from each other for very long."[42] He also made explicit a vague notion that had been floating around in his earlier works. He argued that the descent of languages paralleled the descent of man, that indeed, more primitive animal forms achieved their humanity precisely in acquiring language. As he expressed it: "According to every analogy, man has arisen out of the lower forms, and man, in the proper sense of the word, first became that being when he developed [entwickelten] to the point of language formation."[43] Schleicher further maintained that since human languages were polygenic in origin, so was man. That is, he believed there was no one *Ursprache* whence the other languages descended; rather there were many *Ursprachen*, each having

41. See, for example, August Schleicher, *Die Deutsche Sprache* (Stuttgart: Cotta'scher Verlag, 1860), 58–59.

42. Ibid., 29. From the beginning of his theorizing, Schleicher believed that common *Lautgesetze* (laws of oral expression) governed consonant and vowel changes of language families. In *Deutsche Sprache*, he began formulating macro-laws of language descent, such as the one mentioned previously.

43. Ibid., 38. William Dwight Whitney, commenting on such passages in *Deutsche Sprache*, and comparable ones in *Bedeutung der Sprache*, vigorously dissented: "the rise of language had nothing to do with the growth of man out of an apish stock, but only with his rise out of savagery and barbarism. . . . Man was man before the development of speech began; he did not become man through and by means of it." See Whitney, "Schleicher and the Physical Theory of Language," 324–25.

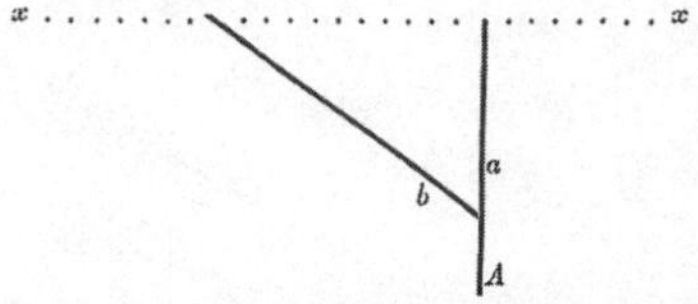


Die Grundsprache A theilt sich in die Sprachen a und b in der beschriebenen Weise nämlich so, daß der Theil des Sprachgebietes b stärkeren Veränderungen unterliegt als der mit a bezeichnete. Bis zum Durchschnitt xx hat also b sich viel weiter von A entfernt als a, und dieß macht eben unser Schema dadurch anschaulich, daß es bx stärker von der geraden Richtung abweichen läßt als ax, das mehr als eine directe Fortsetzung von A erscheint.

FIGURE 8.9 Schleicher's graphic method of showing the greater divergence of daughter language *b* from the mother language *A* and the more lineally descended daughter language *a*. From his *Die Deutsche Sprache* (1860).

developed in different geographical regions out of cries of emotion, imitation, and ejaculation. Since language and thought were two sides of the same process, as language groups developed and evolved independent of one another, so did the different groups of human beings who spoke them.[44]

## SCHLEICHER ON THE EVOLUTION OF MAN, THE LANGUAGE USER

Even before reading Darwin, Schleicher seemed already to have convinced himself that human beings had derived from lower animals. Certainly from the beginning of the nineteenth century, several German biologists—for example, Gottfried Treviranus (1776–1837), Friedrich Tiedemann (1781–1861), and Johann Meckel (1781–1833), all stimulated by Jean-Baptiste de Lamarck (1774–1829)—had become full-blown evolutionists.[45] But was Schleicher full-blown before 1859? His argument for human descent depended on the identification of language with thought. The linkage itself has a venerable history. Authors as far back as Plato understood language and thought to have a close relationship. Johann Gottfried Herder (1744–1803), an author every German intellectual of the first half of the nineteenth century assiduously read, contended, in a prize-

44. Schleicher, *Deutsche Sprache* 5: "Speech is thus the expression of thought in sound, audible thought, just as, on the other hand, thought is inaudible speech."

45. See Robert J. Richards, *Meaning of Evolution: The Morphological Construction and Ideological Reconstruction of Darwin's Theory* (Chicago: University of Chicago Press, 1992), 42–55.

winning treatise of 1772, that language was necessary for thought, "that indeed the first and most elementary application of reason cannot occur without language." Contrary to the creationists, Herder urged that speech arose gradually in human groups, initially through imitation of natural sounds. "No Mercury and Apollo," he protested, "descend from the clouds as by opera machinery—the whole, many-sounding, divine nature is the language teacher and Muse for man."[46] Schleicher would endorse the notion that languages first arose out of imitation of natural sounds, but he conceived an even tighter relationship between language and thought, namely, that of virtual identity.[47] In doing so, he seems proximately to have developed a theoretical position initially laid down by Wilhelm von Humboldt in his *Über die Kawi-Sprache auf der Insel Java* (On the Kawi language on the island of Java, 1836).

In his introduction to the *Kawi-Sprache*—a work often cited by Schleicher—Humboldt argued for the intimate relation between thought and language. He formulated the relationship in this way: "Just as without language no concept is possible, so likewise without language there is no object for the soul, since it is only by means of the concept that any external object can express its complete essence for the soul." Humboldt also suggested, equally darkly, that the descent (*Abstammung*) of language "joined in true and authentic union with physical descent."[48] It would take only slightly more conceptual boldness for Schleicher to conclude, as he forthrightly did, that the descent of language paralleled the descent of thought or mind. Thus the conclusion of *Deutsche Sprache*: with the evolution of different languages comes the evolution of different kinds of human beings.

Yet one can still ask: Did Schleicher's conclusion amount to endorsing something like the Darwinian thesis before Darwin? A clue to the answer to this question can be gleaned from examining a most curious theory in *Deutsche Sprache* concerning the evolution of language in human groups. Schleicher argued that human beings, in their acquisition of language, went through three

46. Johann Gottfried Herder, *Abhandlung über den ursprung der Sprache*, in *Sprachphilosophische Schriften*, ed. Erich Heintel (Hamburg: Felix Meiner, 1975), 3–90, quotations at 28, 32. Michael Forster considers Herder's philosophy of language with considerable dexterity in Forster, "Herder's Philosophy of Language, Interpretation, and Translation: Three Fundamental Principles," in his *After Herder: Philosophy of Language in the German Tradition* (Oxford: Oxford University Press, 2010), 55–90.

47. Thought, according to Schleicher, has material elements—that is, representations (of phenomena) and concepts (when reflexive)—and formal structure—that is, the relationships among the elements. "Language thus has as its task to provide an image in sound of representations and concepts, and their relationships." Meaning (*Bedeutung*) then is the concept or representation as expressed in sound, whereas a word root is the sound complex that expresses meaning. The word itself is the meaning plus the grammatical relationships in sound. See Schleicher, *Deutsche Sprache*, 6.

48. Humboldt, *Kawi-Sprache*, 1:lxxiv, lxxiii.

periods of development: a pre-linguistic period, a prehistorical period of language emergence and evolution, and then a historical period of language decline. In the earliest stage, when no true languages existed, neither did human beings—since without language there could be no human thought. In the next, the prehistorical phase of earth's history, languages (and thus human beings) began to develop. During this period, many different language groups sprang into existence and many died out—indeed, most languages went extinct before achieving their full potential. Others, however, began to spread from one region to another. When languages achieved their maturity, human beings entered the historical period, during which they became self-conscious through the medium of historical understanding. With the advent of the historical period, however, no fundamentally new languages arose. Indeed, during this time, languages began to decline, to devolve. Words started to fall away, forms became simplified, and grammatical relations were lost. Thus Greek and Latin have a much richer store of grammatical forms than do modern languages descended from them. Yet, during this historical period, culture and reason dramatically advanced. Schleicher's scheme of language evolution, with its initial progress and then devolution, seems perfectly paradoxical—that is, until its roots are uncovered.

The fundamental features of this scheme appeared in Schleicher's first monograph, where it is obvious that the basic conception came from Hegel. In *Zur vergleichenden Sprachengeschichte*, Schleicher depicted the three language forms (the isolating, agglutinating, and flexional) as moments in the development of the World Spirit (*Weltgeist*). The Spirit, in the Hegelian view, strove to realize itself, to become fully self-conscious. This striving would be instantiated in the development of human mentality and revealed in language formation. Thus languages would move through dialectical stages, from simple expressions of meaning (in isolating languages), to the structural antithesis in languages that loosely joined meaning and relationships (agglutinating), to a higher synthesis in the "organic unity" of the word, characteristic of the flexional groups—the Semitic and Indo-Germanic. "Whatever we recognize as significant in any sphere of the human spirit," Schleicher averred, "has blossomed from one of these two groups [i.e., Semitic and Indo-Germanic]."[49] In Hegel's view, one explicitly adopted by Schleicher, during the prehistorical period the World Spirit established the intellectual resources—namely, highly developed languages—so as to begin the process of historical self-reflection and

49. Schleicher, *Zur vergleichenden Sprachengeschichte*, 11.

the attainment of freedom. Once the process had begun, however, the energies required for the refined articulation of language began to be employed in the development of rational laws, state governments, and the aesthetic products of advanced civilization. "Hegel thus recognized," according to Schleicher, "the fact that the formation of languages and history cannot take place at the same time, that in the advance of history, rather, language must be worn down."[50]

In Hegel's *Vorlesungen über die Philosophie der Geschichte* (Lectures on the philosophy of history), from which Schleicher initially drew his theory, the prelinguistic period of human existence is represented as nonetheless potentially human, with the "germ or drive" to achieve reflective consciousness already built in.[51] Hegel certainly stopped short of a full-blown biological evolutionism, and this may be where Schleicher himself stopped in *Deutsche Sprache*. Yet, there can be little doubt that Schleicher was brought to the conceptual brink of the theory of biological transformation by Humboldt and Hegel—even if, after 1848, Hegel's name never again appeared in Schleicher's texts. The reading of Darwin's *Origin of Species*, under Haeckel's tutelage, provided the shove for one who was ready to take the plunge into a new conceptual sphere.

Schleicher's own evolutionism obviously went through stages of development, finally resting in his adoption of Darwinism in language and human evolution. One significant index of Darwin's impact on Schleicher's linguistic ideas was the absence of the theory of language decline in his *Darwinsche Theorie*. Darwin's theory of development was thoroughly progressivist; hence it would have been anomalous to suggest that the natural selection of languages led to a devolution of language. Yet Schleicher would have realized that his original assumption of the perfection of ancient languages was one still widely shared by linguists and by cultural critics in love with the classics. He appears to have had only one recourse, which he took—namely, silence. For the most part, however, Darwin's ideas simply overlaid the fundamental features of Schleicher's prior evolutionary project, which derived from the work of those individuals immersed in German Romanticism and Idealism, especially

50. Ibid., 16. Schleicher quoted extensively from Hegel's *Introduction to the Philosophy of History*. This book was part of the compilation of student notes published in 1840, after Hegel's death. Hegel maintained, for instance: "It is a fact, shown by literary remains, that the languages spoken by peoples in uncultured conditions have been well-formed in the highest degree, and that human understanding has developed through having this theoretical foundation. . . . It is further a fact that with the progressive civilizing of society and the state that the systematic activity of the understanding has eroded and language has become less well-formed and poorer." See Georg Wilhelm Friedrich Hegel, *Vorlesungen über die Philosophie der Geschichte*, in *Werke*, 4th ed. (Frankfurt am Main: Suhrkamp, 1995), 12:85.

51. Ibid., 78.

Humboldt and Hegel. They had initially argued that the model of organic growth formed the basic category for understanding the development of consciousness. Their fundamental metaphysical view was monistic—mind and matter expressed two features of an organic *Urstoff*—and this sort of monism became the assumption of evolutionists during the latter half of the nineteenth century, especially of Haeckel.

## HAECKEL'S THEORY OF THE LINGUISTIC EVOLUTION OF MAN

Ernst Haeckel, to whom Schleicher's *Darwinsche Theorie* had been addressed, himself had converted to Darwinism in 1860, virtually as soon as he read the German translation of the *Origin*.[52] At the time, he was working on his habilitation, in which he would describe and systematically classify the radiolaria, simple one-celled creatures that inhabited the oceans and exuded an exoskeleton. Darwin's theory helped him make sense of the myriad families, genera, and species these creatures displayed.[53] Haeckel, like Schleicher, had been ready for such a theory as Darwin's; he too was thoroughly imbued with Romantic ideals. His letters to his fiancée—written while working on his habilitation in southern Italy—are smeared with quotations from Goethe. The romantic élan so took his soul in thrall that he contemplated giving up his scientific work for that of the life of a painter and free spirit. For a time he wandered over the island of Capri with a poet friend, who almost seduced him, quite literally, away from his eventual career as a university professor. It was only the thought of his fiancée, with whom he was deeply in love, and the realization that the life of a Bohemian did not pay very well that steeled him to finish his habilitation and return to Jena.

Haeckel remained at Jena throughout his career. Under his influence during the last half of the nineteenth century, the university became a bastion of Darwinian thought. Schleicher, who quickly slid to the Darwinian side under his friend's guidance, in turn contributed to Haeckel's own version of Darwinism, a version that became part of the standard view through the early years of the twentieth century. Schleicher made several significant contributions. First, he confirmed, from a quite different perspective, Darwin's theory, and thus sup-

52. See chapter 6 in this volume.

53. Ernst Haeckel, *Die Radiolarien. (Rhizopodia radiaria). Eine Monographie*, 2 vols. (Berlin: Reimer, 1862). See chapter 6 in the present volume.

ported Haeckel in what would become a comprehensive scientific philosophy. Second, he solidified for his friend the important metaphysical vision that became the basis for evolutionary theory in the latter half of the nineteenth century, namely, monism.

Monism could support a variety of philosophical refinements. For instance, the American pragmatists William James (1842–1910) and John Dewey (1859–1952) both avowed monism. Henri Bergson (1859–1941) also claimed that metaphysical doctrine, as did most other evolutionists. Haeckel himself elevated the doctrine into a "monistic religion," as he termed it.[54] The philosophy of monism could be given, as the works of these individuals suggest, different spins, different emphases. Haeckel always reminded his readers that anything called *Geist* had a material side. So, for example, under the rubric of monism in his *Natürliche Schöpfungsgeschichte* (The natural history of creation, 1868)—which was a popular version of his *Generelle morphologie der organismen* (The general morphology of organisms, 1866), his fundamental theoretical work—Haeckel insisted that "the human soul has been gradually formed through a long and slow process of differentiation and perfection out of the vertebrate soul." Or, as he also put it: "Between the most highly developed animal soul and the least developed human soul there is only a quantitative, but no qualitative difference." Indeed, Haeckel thought that the mental divide separating the lowest man (the Australian or Bushman) and the highest animal (ape, dog, or elephant) was smaller than that separating the lowest man from the highest man, a Newton, a Kant, or a Goethe.[55] Haeckel regarded differences among men as so significant that he thought humankind should be classified not simply into different races or varieties of one species, but into some nine separate species of one genus (see fig. 9.2).

Morphological similarities led Haeckel to argue that human beings evolved through a kind of bottleneck, that of the narrow-nosed apes (fig. 8.10). There must have been, according to Haeckel, an *Urmensch*, or *Affenmensch*—an ape-man—which stemmed from the *Menschenaffen*—the menlike apes. This was the missing link, and we owe the currency of this idea to Haeckel. He thought

54. See Ernst Haeckel, *Der Monismus als Band zwischen Religion und Wissenschaft* (Bonn: Emil Strauss, 1892). Haeckel first explicitly endorsed Schleicher's conception of monism in Haeckel, *Generelle morphologie der organismen*, 2 vols. (Berlin: Reimer, 1866), 1:105–8.

55. Ernst Haeckel, *Die Natürliche Schöpfungsgeschichte* (Berlin: Reimer, 1868), 550 and 546 (quotations), 549.

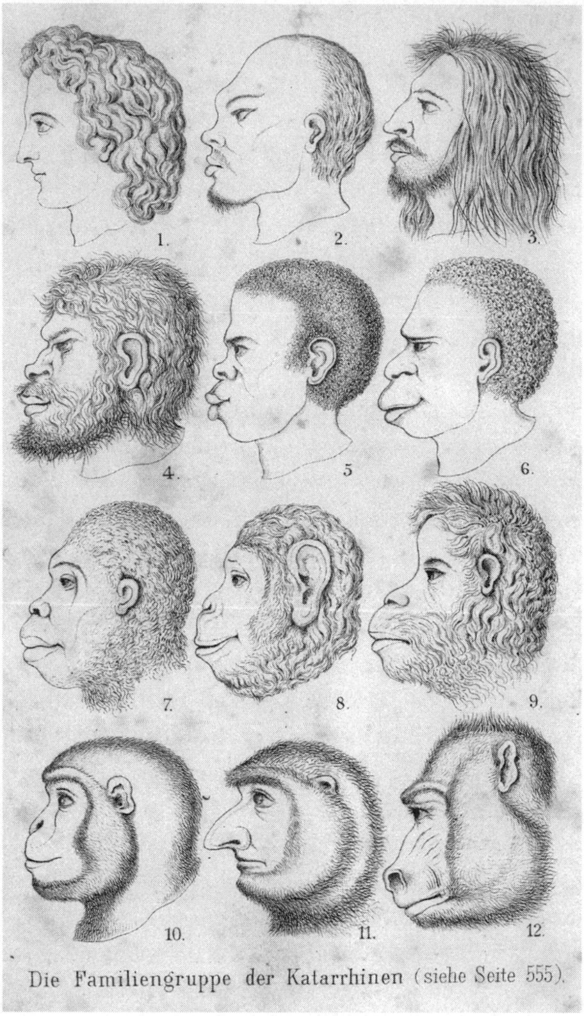


FIGURE 8.10 Frontispiece showing the descent of the several human species from the narrow-nosed apes. From Ernst Haeckel, *Natürliche Schöpfungsgeschichte* (1868).

the *Affenmensch* would likely have come either from Africa or perhaps from the area of the Dutch East Indies, where the orangutan was to be found. Later, Haeckel would name this Ur-ancestor *Pithecanthropus alalus*—ape-man without speech. His protégé Eugene Dubois (1858–1940), a Dutch army doctor, actually found Pithecanthropus in Java in 1891, and the missing link, which

Haeckel had predicted, became widely celebrated.[56] It was later rechristened *Homo erectus*, and Java man was the first of his remains to be discovered.

The unspoken question about human evolution, for which Haeckel had a spoken answer, was this: What essentially distinguished the various species of men, what led to this great mental differentiation—a differentiation that persuaded him that the Papuan, for instance, was intellectually closer to the apes than to a Newton or Goethe? Morphologically, after all, aside from skin color and hair differences, human beings were pretty much alike. On this question Schleicher made another contribution. The monistic metaphysics that he professed emphasized the mental side of things, which is not surprising given his early commitment to Romantic Idealism. In *Zur vergleichenden Sprachengeschichte*, he argued, in Hegelian fashion, that the systematic representation of beings, from the logically simple to the more complex, was identical to the becoming of those beings in time, in a kind of evolutionary emanation. Animal cognition, in this philosophical consideration, remained decisively different from human mentality. By the 1860s, Schleicher would ground his philosophical conception on a scientifically articulated one, namely, Darwin's. Even in the 1860s, however, he still maintained that human beings were quite distinct from animals in their mental ability. Human mentality was exhibited in language, of which no animal was capable. What this now meant, however, was that the advent of language created man out of his apelike forbearers, a creation that would not be repeated. Since, according to Schleicher, the basic language groups did not evolve from one another, each protohuman group became human in a distinctively different way. After the initial establishment of the isolating, agglutinating, and flexional languages, which created the different groups of human beings, these language animals evolved at different rates and in different directions. Only the Indo-Germanic and Semitic languages reached a kind of perfection not realized in the other groups. Here, then, was Haeckel's solution to the evolution of the various human species.

In the *Natürliche Schöpfungsgeschichte*, Haeckel maintained that human beings had a quasi-monogenic origin in *Pithecanthropus*. He imagined that these original protohumans evolved on a continent that now lay sunken in the Indian Ocean, somewhere between Malay and South West Africa, and that these primitive *Urmenschen* eventually split into two groups, which migrated, respectively, toward east and west. Later he would call this fanciful continent "Atlantis" or "Paradise," with the full irony of that latter name in mind.

56. See Eugene Dubois, *Pithecanthropus erectus: Eine Menschenaenliche Uebergangsform aus Java* (Batavia: Landsdruckeri, 1894).

Although our physical frame could be traced back to this one kind of ape-man, Haeckel yet maintained that, in a proper sense, the human species were polygenic, as Schleicher had suggested:

> We must mention here one of the most important results of the comparative study of languages, which for the *Stammbaum* of the species of men is of the highest significance, namely that human languages probably had a multiple or polyphyletic origin. Human language as such probably developed only after the species of speechless *Urmenschen* or *Affenmenschen* had split into several species or kinds. With each of these human species, language developed on its own and independently of the others. At least this is the view of Schleicher, one of the foremost authorities on this subject. . . . If one views the origin of the branches of language as the special and principal act of becoming human, and the species of humankind as distinguished according to their language stem, then one can say that the different species of men arose independently of one another.[57]

The clear inference is that the languages with the most potential created the human species with the most potential. And, as Haeckel never tired of indicating, that species with the most potential—a potential realized—was constituted by the Semitic and Indo-Germanic groups, with the Berber, Jewish, Greco-Roman, and Germanic varieties in the forefront.[58] Their vertical position on the human *Stammbaum* indicated the degree of their evolutionary advance (see fig. 9.2).

Schleicher's greatest and lasting contribution to evolutionary understanding may simply be his use of a *Stammbaum* to illustrate the descent of languages. Not long after Schleicher published his open letter, Haeckel finished his magnum opus, his synthesis of evolutionary theory and morphology, the large two-volume *Generelle Morphologie der Organismen*. The end of the second volume included eight tables of phylogenetic trees. While there are some vague antecedents for the graphic use of treelike forms for the expression of descent relationships, Haeckel obviously took his inspiration from his good

57. Haeckel, *Natürliche Schöpfungsgeschichte*, 511.

58. The debate over the monogenic or polygenic origin of man still rages, if in a slightly different key. See, for instance, Christopher Stringer and Robin McKie, *African Exodus: The Origins of Modern Humanity* (New York: Holt, 1996). See also my review of their book, Richards, "Neanderthals Need Not Apply," *New York Times Book Review*, 17 August 1997, 10.

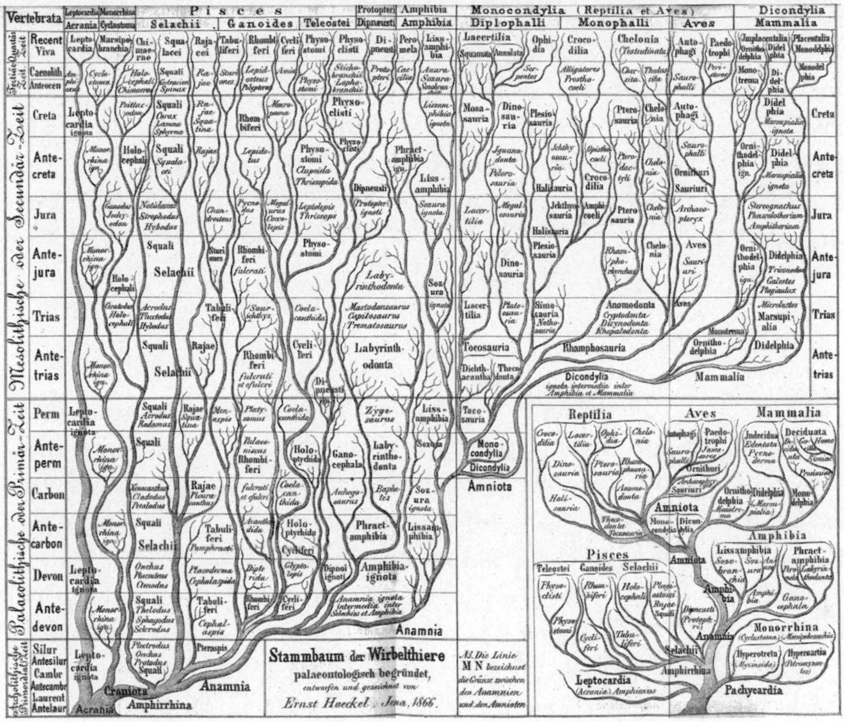


FIGURE 8.11 Haeckel's stem-tree of the descent of vertebrates. From his *Generelle Morphologie der Organismen* (1866).

friend Schleicher. And Haeckel's *Stammbäume* have become models for the representation of descent ever since.

Haeckel's tree of vertebrates (fig. 8.11) may be compared with both Darwin's diagram and Schleicher's. Unlike Darwin's and but like Schleicher's, Haeckel's illustration shows a single origin of the vertebrate phylum, though each of the major phyla (e.g., mollusca, articulata, etc.), he maintained, had independent origins. And, of course, again unlike Darwin's but like Schleicher's, Haeckel's *Stammbaum* depicts actual species, the extinct and the extant. Schleicher's tree captured both time, marked as the distance from the Indo-Germanic *Ursprache*, and morphological differentiation, represented by the separation of the branches. This too Haeckel's diagram depicts. Haeckel's tree has an added feature, of course: it actually looks like a tree, whereas Darwin's and Schleicher's sketches are merely line drawings. This may seem, at bottom, a trivial difference, arising from the fact that Haeckel was an accomplished artist. Certainly his talent made the depiction possible. But the living, branching, gnarled, German oak functioned as a kind of graphic rhetoric: it vividly displayed the tree of life, in all its gothic and romantic textures. In the case of all three authors, but with increasing vivacity, a visual argument was made, which with Haeckel became a powerful, if silent, linking of the very newest theory in biology with the traditions of German Romanticism well established at Jena.

## CONCLUSION

During the mid-1860s, Darwin's great friend Wallace had developed some powerful arguments to show natural selection to be insufficient to account for man's big brain.[59] Darwin saw the force of his friend's argument, and thus the vexing problem it posed—How to explain the complex mind and big brain of human beings? But during the mid-1860s, another kind of argument came to his attention, through several related sources. The argument was Schleicher's for the linguistic creation of man.

Darwin studied Schleicher's *Darwinsche Theorie*, which he then used and cited in his own account of human evolution in the *Descent of Man*. He got two other doses of Schleicher's views more indirectly. Frederick Farrar—whom Darwin named along with his cousin Hensleigh Wedgwood and Schleicher as contributing to his conception of language—had made Schleicher's theories known to the British intellectual community through a comprehensive account in the journal *Nature*.[60] Schleicher's conceptions also got conveyed to Darwin through a gift of Haeckel's *Natürliche Schöpfungsgeschichte*, which the author sent in 1868. Darwin wrote to a friend after reading Haeckel's work that it was "one of the most remarkable books of our time."[61] Darwin's notes and underlining in the book are extensive. He was particularly interested, as shown by his scorings and marginalia, in Haeckel's account of Schleicher's thesis, as set out by Schleicher in *Über die Bedeutung der Sprache für die Naturgeschichte des Menschen*. Here,

59. Wallace first advanced his arguments in a review of new editions of Charles Lyell's works. See Alfred Russel Wallace, "Review of *Principles of Geology* by Charles Lyell; *Elements of Geology* by Charles Lyell," *Quarterly Review* 126 (1869): 359–94. See chapter 4 of the present volume.

60. Farrar, "Philology and Darwinism."

61. Darwin to William S. Dallas (9 June 1868), in *The Correspondence of Charles Darwin*, ed. Frederick Burkhardt et al., 19 vols. to date (Cambridge: Cambridge University Press, 1985–),16 (pt. 1): 573.

then, Darwin had a counterargument to Wallace's, one by which he could solidify an evolutionary naturalism: language might modify brain, increasing its size and complexity, and that enlargement might become a permanent, hereditary legacy. The historical irony, of course, is that Darwin's evolutionary naturalism obtained its support, through Schleicher, ultimately from Wilhelm von Humboldt and Georg Friedrich Hegel, two foremost representatives of German Romanticism and Idealism. The German Romantics and Idealists thus forged the missing link in nineteenth-century evolutionary theory.

CHAPTER NINE

# Was Hitler a Darwinian?

*The Darwinian underpinnings of Nazi racial ideology are patently obvious. Hitler's chapter on "Nation and Race" in* Mein Kampf *discusses the racial struggle for existence in clear Darwinian terms.*

—Richard Weikart, historian, California State University, Stanislaus[1]

*Hamlet: Do you see yonder cloud that's almost in shape of a camel?*

—Shakespeare, *Hamlet* III.ii.2

Several scholars and many religious conservative thinkers have recently charged that Hitler's ideas about race and racial struggle derived from the theories of Charles Darwin (1809–1882), either directly or through intermediate sources. For example, the historian Richard Weikart, in his book *From Darwin to Hitler*, maintains: "No matter how crooked the road was from Darwin to Hitler, clearly Darwinism and eugenics smoothed the path for Nazi ideology, especially for the Nazi stress on expansion, war, racial struggle, and racial extermination."[2] In a subsequent book, *Hitler's Ethic: The Nazi Pursuit of Evolutionary Progress*, Weikart argues that Darwin's "evolutionary ethics drove him

1. Richard Weikart, "Was It Immoral for *Expelled* to Connect Darwinism and Nazi Racism?" Discovery Institute (http://www.discovery.org/a/5069).

2. Richard Weikart, *From Darwin to Hitler: Evolutionary Ethics, Eugenics, and Racism in Germany* (New York: Palgrave Macmillan, 2004), 6.

[Hitler] to engage in behavior that the rest of us consider abominable."[3] The epigram to this chapter makes Weikart's claim patent. Other critics have also attempted to forge a strong link between Darwin's theory and Hitler's biological notions. In the 2008 documentary film *Expelled*, a defense of Intelligent Design, the Princeton-trained philosopher David Berlinski, in conversation with Weikart, confidently asserts: "If you open *Mein Kampf* and read it, especially if you can read it in German, the correspondence between Darwinian ideas and Nazi ideas just leaps from the page."[4] John Gray, former professor at the London School of Economics, does allow that Hitler's Darwinism was "vulgar."[5] Hannah Arendt also appears to have endorsed the connection when she declared: "Underlying the Nazis' belief in race laws as the expression of the law of nature in man, is Darwin's idea of man as the product of a natural development which does not necessarily stop with the present species of human being."[6] Even the astute historian Peter Bowler comes close to suggesting a causal connection between Darwin's accomplishment and Hitler's: "By making death a creative force in nature . . . Darwin may indeed have unwittingly helped to unleash the whirlwind of hatred that is so often associated with his name."[7] Put "Darwin and Hitler" in a search engine and hundreds of thousands of hits will be returned, most from religiously and politically conservative websites, articles, and books.

With the exception of the aforementioned, most scholars of Hitler's reign don't argue for a strong link between Darwin's biology and Hitler's racism, but they often deploy the vague concept of social Darwinism when characterizing Hitler's racial ideology.[8] The very name of the concept—whatever its content—does suggest a link with evolutionary theory and particularly Darwin's version

3. Richard Weikart, *Hitler's Ethic: The Nazi Pursuit of Evolutionary Progress* (New York: Palgrave Macmillan, 2009), 2–3.

4. *Expelled: No Intelligence Allowed* (Rocky Mountain Pictures, 2008), a documentary film written by Kevin Miller and Ben Stein and directed by Nathan Frankowski. The line by Berlinski comes sixty-four minutes into the film.

5. John Gray, "The Atheist Delusion," *Guardian*, 15 March 2008, 4.

6. Hannah Arendt, *The Origins of Totalitarianism* (Orlando, FL: Harcourt, [1948] 1994), 463.

7. Peter Bowler, "What Darwin Disturbed: The Biology That Might Have Been," *Isis* 99 (2008): 560–67, quotation at 564–65.

8. Here are a few of the more recent scholars who have described Hitler as a social Darwinist: Joachim Fest, *Hitler*, trans. Richard Winston and Clara Winston (New York: Harcourt Brace Jovanovich, 1974), 54–56; Mike Hawkins, *Social Darwinism in European and American Thought, 1860–1945* (Cambridge: Cambridge University Press, 1997), 277–78; David Welch, *Hitler* (London: Taylor and Francis, 1998), 13–15; Frank McDonough, *Hitler and the Rise of the Nazi Party* (London: Pearson/Longman, 2003), 5; Richard Evans, *The Coming of the Third Reich* (New York: Penguin, 2003), 34–37; and Stephen Lee, *Hitler and Nazi Germany* (London: Rutledge, 2010), 94.

of that theory. The supposed connection between Darwin's conceptions and Hitler's is often traced through the biological ideas of the English scientist's German disciple and friend, Ernst Haeckel (1834–1919).

In his book *The Scientific Origins of National Socialism* (1971), Daniel Gasman claimed: "Haeckel . . . was largely responsible for forging the bonds between academic science and racism in Germany in the later decades of the nineteenth century."[9] In a more recent book, Gasman urged that Haeckel had virtually begun the work of the Nazis: "For Haeckel, the Jews were the original source of the decadence and morbidity of the modern world and he sought their immediate exclusion from contemporary life and society."[10] Gasman's judgment received the imprimatur of Stephen Jay Gould, who concluded in his *Ontogeny and Phylogeny* (1977): "But as Gasman argues, Haeckel's greatest influence was, ultimately, in another tragic direction—National Socialism. His evolutionary racism; his call to the German people for racial purity and unflinching devotion to a 'just' state; his belief that harsh, inexorable laws of evolution ruled human civilization and nature alike, conferring upon favored races the right to dominate others; the irrational mysticism that had always stood in strange communion with his grave words about objective science—all contributed to the rise of Nazism."[11]

Scholars such as Gould, Bowler, and Larry Arnhart—as well as a host of others—attempt to distinguish Haeckel's views from Darwin's so as to exonerate the latter while sacrificing the former to the presumption of a strong causal connection with Hitler's anti-Semitism.[12] I don't believe this effort to disengage Darwin from Haeckel can be easily accomplished, since on central matters—descent of species, struggle for existence, natural selection, inheritance of acquired characters, recapitulation theory, progressivism, hierarchy of races—no essential differences between master and disciple exist.[13] So if Hitler endorsed Haeckel's evolutionary ideas, he thereby also endorsed Darwin's.

9. Daniel Gasman, *The Scientific Origins of National Socialism: Social Darwinism in Ernst Haeckel and the German Monist League* (New York: Science History Publications, 1971), 40.

10. Daniel Gasman, *Haeckel's Monism and the Birth of Fascist Ideology* (New York: Peter Lang, 1998), 26.

11. Stephen Jay Gould, *Ontogeny and Phylogeny* (Cambridge: Harvard University Press, 1977), 77–78.

12. See Peter Bowler, *The Non-Darwinian Revolution: Reinterpreting a Historical Myth* (Baltimore: Johns Hopkins University Press, 1988), 83–84; and Larry Arnhart, *Darwinian Conservatism* (Charlottesville, VA: Imprint Academic, 2005), 116.

13. I have shown the essential identity of Darwin's and Haeckel's evolutionary theories at some length in Robert J. Richards, *The Tragic Sense of Life: Ernst Haeckel and the Struggle over Evolutionary Thought* (Chicago: University of Chicago Press, 2008), 135–62. Gliboff also argues that although some scholars have contrasted Darwin's and Haeckel's views on morphological type, their theories were basically the

## THE SUPPOSED CAUSAL CONNECTION BETWEEN DARWIN AND HITLER

Those critics who have urged a conceptually causal connection between Darwin's or Haeckel's biology and Hitler's racial beliefs—Weikart, Berlinski, and a myriad of religiously and politically constricted thinkers—apparently intend to undermine the validity of Darwinian evolutionary theory and, by regressive implication, morally indict Darwin and Darwinians like Ernst Haeckel. More reputable scholars—Gould, Arnhart, Bowler, and numerous others—are willing to offer up Haeckel to save Darwin by claiming significant differences between their views, a claim, as I've suggested, that cannot be sustained. The arguments arrayed against Darwin and Haeckel have power, no doubt. Whether they *should* have power is the question I investigate here.

Two salient issues arise out of the allegations of a connection between Darwinian theory and Hitler's racial conceptions: first, the factual truth of the claimed causal connections; and second, the epistemic and moral logic that draws implications from the supposed connections. The factual question can be considered at four levels. These distinctions may seem tedious to the impatient, but they are necessary, since the factual claim is often settled by even talented scholars through the deployment of a few vague observations. First, there is the epistemological problem of the very meaning of the assertion of causal connections among ideas. This issue falls under the rubric of *influence*, that is, one individual's ideas influencing or having causal impact on those of another. A host of acute epistemological problems attend the conception of influence (ideas, after all, are not like billiard balls), but I bracket them in this discussion and simply assume that influence is real and causally potent. The second level of the factual question is this: Did Hitler embrace Darwinian theory? Third, did any supposed endorsement actually lead to his racial policies, especially concerning the treatment of Jews? Finally, we should consider the beliefs and attitudes of those scientists working directly under the authority of the Nazi party: Did they adopt Darwinian theory and on that basis urge the inferiority of Jews and recommend eugenic measures? I will consider each of these latter three levels of the factual question in turn.

There is a kind of pseudo-historical game that can be played with causal influence, a distraction that will vitiate a serious attempt to deal with the second and third levels of the factual question. Instead of tracing out a reputed serious

same. See Sander Gliboff, *H. G. Bronn, Ernst Haeckel, and the Origins of German Darwinism* (Cambridge: MIT Press, 2008), 161–66.

engagement by Hitler with Darwin's ideas and making an effort to determine how those ideas might have actually motivated him, one could play something like "Six Degrees of Charles Darwin." That is, one could catch Hitler using, say, a certain phrase he picked up from someone whom he'd read, who in turn had read someone else who used the phrase, who found it in a journal article that mentioned someone quoting Darwin, and so forth. Virtually any remarks made by Hitler could thus be traced back to Darwin—or to Aristotle, or to Christ. The real issue would be whether the phrase had Darwinian ideas behind it and whether such usage by Hitler motivated his actions.

The proposition that Darwinian ideas motivated Hitler's anti-Jewish racism moves quickly to the edge of profound absurdity without the need of any scholarly pressure. As Hugh Trevor-Roper argued long ago, Hitler failed to establish a coherent, central administrative power in the Nazi state; rather, he allowed individual factions within the government to gather resources in greed and control them in fear. He waged total war without any general strategy.[14] But more disastrously, he eliminated that portion of the population—replete with technical expertise in management, business, and sciences—that could have provided the margin of ultimate victory. And he knew the Jews had such talent, even if they lacked, in his estimation, the requisite culture. He wasted his most valuable resource and expended manpower and money in doing so. No abstract scientific theory could have motivated such irrationality. At the very best, he might have used some fugitive phrases to disguise the mania that really drove him. But, for the moment, I will suspend this objection, for as I will show, not even his suspect language has a Darwinian provenance.

Attendant on the factual question is that of the meaning of social Darwinism when applied to Hitler and other Nazis. The term is maddeningly opaque, but we can discriminate several different notes that conventionally fall under the conception and then decide which of those notes apply to the Nazis, and to Hitler in particular.

The strategy of those attempting to show a causal link between Darwin's theory and Hitlerian ideas about race runs, I believe, like this: the causal relation of influence proceeding from Darwin to future Nazi malevolence justifies regressive epistemic and moral judgments running from the future back to the past, thus indicting Darwin and individuals like Haeckel with moral responsibility for the crimes of Hitler and his minions and thereby undermining evolutionary theory. Now the validity of this kind of moral logic might be

14. Hugh Trevor-Roper, *The Last Days of Hitler*, 6th ed. (Chicago: University of Chicago Press, 1992), 53–90.

dealt with straightaway: even if Hitler had the *Origin of Species* as his bedtime reading and clearly derived inspiration from it, this would have no bearing on the truth of Darwin's theory or directly on the moral character of Darwin and other Darwinians. Mendelian genetics became ubiquitous as a scientific foundation for Nazi eugenic policy (and American eugenic proposals as well), though none of the critics questions the basic validity of that genetic theory or impugns Mendel's moral integrity. Presumably Hitler and other party officials recognized chemistry as a science and utilized its principles to exterminate efficiently millions of people. But this hardly precludes the truth of chemical theory or morally taints all chemists. It can only be rampant ideological confusion to maintain that the alleged connection between Hitler's ideas and those of Darwin and Haeckel, ipso facto, nullifies the truth of evolutionary theory or renders these evolutionists, both long dead before the rise of the Nazis, morally responsible for the Holocaust.

If Hitler and leading Nazi biologists had adopted Darwinian theory, exactly what feature of the theory would supposedly have induced them to engage in morally despicable acts? Weikart, for one, asserts that it was Darwinian materialism that "undercut Judeo-Christian ethics and the right to life."[15] This charge has three salient problems. First, strictly speaking, Darwin was not a materialist; when the *Origin* was published he was a theist.[16] The leading Darwinian in Germany in the late nineteenth century, Ernst Haeckel, rejected the charge of materialism; he was a convinced Goethean monist (i.e., all organisms had a material side and a mental side). It is true, however, that Darwin and Haeckel were perceived as materialists by many later critics—and by historians like Weikart. Second, as I'll indicate in a moment, Darwin's own moral theory did not abandon Judeo-Christian precepts. Nor did Haeckel's. Haeckel was quite clear. He accepted the usual moral canon: "Doubtless, human culture today owes the greater part of its perfection to the spread and ennobling effect of Christian ethics."[17] Haeckel, like Darwin, simply thought that Christian precepts had a source other than Divine command; those norms derived from the altruism bred in the bone by natural selection.[18] But the chief reason why

15. Richard Weikart, "Darwinism and Death: Devaluing Human Life in Germany, 1859–1920," *Journal of the History of Ideas* 63 (2002): 323–44, quotation at 343.

16. Charles Darwin, *The Autobiography of Charles Darwin, 1809–1882*, ed. Nora Barlow (New York: Norton, 1969), 92–93. Only in the mid-1860s did Darwin's theism slip away; he constructed his theory as a theist. See chapter 2 in the present volume.

17. Ernst Haeckel, *Der Monismus als Band zwischen Religion und Wissenschaft* (Bonn: Emil Strauss, 1892), 29.

18. I have discussed Haeckel's ethical position in Richards, *Tragic Sense of Life*, 352–54.

presumptive Darwinian materialism cannot be the source of the malign actions of Hitler and leading Nazi biologists is simple: they were not materialists. As I show later in the chapter, Hitler's gauzy mystical attitude about *Deutschtum* and the German race was hardly materialistic; moreover, leading Nazi biologists rejected Darwin and Haeckel precisely because the theories of these two scientists were, it was thought, materialistic, while *volkisch* biology was not. In the first instance, however, it is crushingly naïve to believe that an extremely abstract metaphysical position, such as materialism—or vitalism—can distinctively produce morally deleterious or virtuous behavior. In this instance, though, whether abstract ethereal belief or not, Darwinian theory cannot be the root of any malign influence perpetrated on the Nazis for the reason Weikart asserts. Below I will describe the character of the more rarified metaphysics of Nazi scientists to show why it had no connection with Darwinism. Another consideration further attenuates the gossamer logic of the arguments mounted by Weikart, Berlinski, Gasman, Gould, and members of the Intelligent Design crowd: their exclusive focus on the supposed Darwin–Hitler or Haeckel–Hitler connection reduces the complex motivations of the Nazi leaders to linear simplicity.

The critics I have mentioned, and many others besides, ignore the economic, political, and social forces operative in Germany in the 1920s and 1930s, and they give no due weight to the deeply rooted anti-Semitism that ran back to Luther and medieval Christianity and forward to the religious and political sentiments rife at the end of the nineteenth century.[19] The names of those who prepared the ground before Hitler entered the scene go unmentioned: the court preacher and founder of the Christian Socialist Party, Adolf Stöcker (1835–1909), who thought the Jews threatened the life-spirit of Germany; Wilhelm Marr (1819–1904), founder of the League of Anti-Semitism, who maintained that the Jews were in a cultural "struggle for existence" with the spirit of Germanism, taking over the press, the arts, and industrial production; or the widely read historian Heinrich von Treitschke (1834–1896), who salted his historical fields with animadversions about alien Jewish influences on German life and provided the Nazis with the bywords "the Jews are our misfortune."[20]

19. Richard Evans discusses this mix of religious and political anti-Semitism at the end of the nineteenth century in Evans, *Coming of the Third Reich*, 22–34.

20. See, for example, Adolf Stöcker, *Das modern Judenthum in Deutschland besonders in Berlin* (Berlin: Verlag von Wiegandt und Grieben, 1880), 4: "the entire misery of Germany, I should have mentioned, comes from the Jews." See also Wilhelm Marr, *Der Sieg des Judenthums über das Germanenthum, vom nicht confessionellen Standpunkt aus betrachtet*, 8th ed. (Bern: Rudolph Costenoble, 1879). Marr held that "the degradation of the German state to the advantage of Jewish interests is a goal pursued everywhere. The daily press is chiefly in Jewish hands and they have made a speculative and industrial matter out of

FIGURE 9.1 Richard Wagner (1813–1883), in 1881.
(© National Portrait Gallery, London)

Then there was the composer Richard Wagner (1813–1883; fig. 9.1), whose music Hitler adored, even as a young man attending countless performances of *The Flying Dutchman*, *Parsifal*, *Lohengrin*, and the Ring cycle, and as rising political leader visiting the maestro's home in Bayreuth at the invitation of the

journalism, a business forming public opinion—theater criticism, art criticism are three-quarters in Jewish hands. . . . There is no 'struggle for existence,' except that Judaism gathers its advantage" (24, 27). See also Heinrich von Treitschke, *Ein Wort über unser Judenthum* (Berlin: G. Reimer, 1880), 4: "ertönt es heute wie aus einem Munde: 'die Juden sind unser Unglück!'"

Wagner family. In 1850 Wagner wrote a small pamphlet, which he reissued and expanded in 1869, titled *Das Judenthum in der Musik* (Jewishness in music). He wished "to explain the involuntary revulsion we have for the personality and nature of the Jews and to justify this instinctive repugnance, which we clearly recognize and which is stronger and more overwhelming than our conscious effort to rid ourselves of it."[21] These are only a few of the intellectuals—or near-intellectuals—who expressed unreflective to more consciously aggressive anti-Semitic attitudes at the turn of the century; their malevolent depictions and vicious rants cascaded through German intellectual society in the early years of the twentieth century. Of course, these attitudes were not confined to Germany but invaded distant shores as well. The new U.S. ambassador to Germany in 1933, William E. Dodd (1869–1940), former chair of the history department of which I am currently a member, could, for example, discount the outrageous attacks on Jews in Berlin by SA troops with the casual remark to a Nazi official that "we have had difficulty now and then in the United States with Jews who had gotten too much of a hold on certain departments of intellectual and business life."[22] Dodd finally came to appreciate that the Nazi treatment of Jews went beyond the bounds of "civilized" anti-Semitism, and he became an early voice of warning about the intentions of Hitler's government. The disposition of Dodd and the others I have just mentioned were innocent of any concern with Darwin's theory. Finally, one needs consider the politicians, especially in Vienna, who used anti-Semitism in opportunistic ways. I will examine the views of these figures more particularly later in the chapter, since Hitler himself ascribed his racial attitudes to this source. The critics of Darwin and Haeckel have in their indictments neglected the various complex social and cultural forces that fueled the anti-Semitic obsessions of Hitler and his henchmen. The critics have sought, rather, to discover a unique key to Nazi evil.[23]

The presumption that a factual connection between Darwin's *Origin of Species* and Hitler's *Mein Kampf* morally indicts Darwin and somehow undermines evolutionary theory rests, quite obviously, on defective moral and epistemic logic—rather, on no logic at all. Nonetheless, I put aside this logical consideration for the moment to investigate the supposed factual linkage.

21. Richard Wagner, *Das Judenthum in der Musik* (Leipzig: Weber, 1869), quotation at 10–11.

22. Quoted by Erik Larson, *In the Garden of Beasts* (New York: Crown Books, 2011), 130.

23. Despite the caveats I've offered about the easy slide from causal influence to epistemic and moral indictment, I don't want to deny that under certain well-defined circumstances one might justify, for instance, a morally negative assessment based on a relationship of conceptual influence. I have analyzed those circumstances in Robert J. Richards, "The Moral Grammar of Narratives in History of Biology—The Case of Haeckel and Nazi Biology," in *The Cambridge Companion to the Philosophy of Biology*, ed. Michael Ruse and David Hull (Cambridge: Cambridge University Press, 2007), 429–52.

## DARWINIAN THEORY AND RACIAL HIERARCHY

The first factual issue to tackle is this: Did Hitler embrace Darwinian theory? The question, however, needs to be made more exact: What features of Darwin's theory did he embrace, if any? Concerning the theory, especially as applied to human beings, we can discriminate three central components: (1) that human groups can be arranged in a racial hierarchy from less advanced to more advanced; (2) that species have undergone descent with modification over vast stretches of time and that human beings, in particular, descended from apelike ancestors; and (3) that natural selection is the principal device to explain species transitions. Now the questions become: Did Hitler adopt any of these positions, and were they derived ultimately from Darwin? And did these ideas cause him to adopt or favor racist and specifically anti-Semitic views characteristic of Nazi biology? Of course, a positive answer to this latter question is essential to complete the causal connection between Darwinian theory and Hitler's lethal racial attitudes.

The first component of Darwinian theory to consider is that of racial hierarchy. Gould argued that Darwin's theory was not progressivist, and therefore it did not situate species and races, particularly the human races, in any hierarchical scheme. He maintained, for example, that "an explicit denial of innate progression is the most characteristic feature separating Darwin's theory of natural selection from other nineteenth-century evolutionary theories."[24] Lamarck, by contrast, had postulated an internal, quasi-hydraulic mechanism that produced progressively more complex species over time. And Haeckel, quite graphically, arranged the human groups in a hierarchical scheme. Although other scholars have followed Gould's lead,[25] it is clear that Darwin thought of natural selection as a kind of external force that would generally produce, over vast stretches of time, more progressively developed organisms. In the penultimate paragraph of the *Origin of Species*, he explicitly stated his

24. Stephen Jay Gould, "Eternal Metaphors of Palaeontology," in *Patterns of Evolution as Illustrated in the Fossil Record*, ed. A. Hallan (New York: Elsevier, 1977), 1–26, quotation at 13. Gould subsequently tried to distinguish between what Darwin's theory demanded and what his cultural dispositions might have led him to assert—as if Darwin's theory were not embedded in the words of his books. See Stephen Jay Gould, *Wonderful Life: The Burgess Shale and the Nature of History* (New York: Morton, 1989), 257–58. I have discussed Darwin's progressivism vis-à-vis the assertions of Gould, Peter Bowler, and Michael Ruse. See Robert J. Richards, "The Epistemology of Historical Interpretation," in *Biology and Epistemology*, ed. Richard Creath and Jane Maienschein (Cambridge: Cambridge University Press, 2000), 64–90.

25. See, for example, Peter Bowler, *Theories of Human Evolution* (Baltimore: Johns Hopkins University Press, 1986), 13.

view: "And as natural selection works solely by and for the good of each being, all corporeal and mental endowments will tend to progress toward perfection."[26] Even before he formulated his theory, however, Darwin was disposed to regard certain races as morally and intellectually inferior, as, for example, the Fuegian Indians he encountered on the *Beagle* voyage. His later theoretical formulations and his own cultural assumptions surely reinforced each other. In the *Descent of Man*, Darwin described the races as forming an obvious hierarchy of intelligence and moral capacity, from savage to civilized, with the "intellectual and social faculties" of the lower races comparable to those that must have characterized ancient European man. Accordingly, he ventured that "the grade of their civilisation seems to be a most important element in the success of competing nations," which explained for him the extermination of the Tasmanians and the severe decline in population of the Australians, Hawaiians, and Maoris.[27] Those groups succumbed in the struggle with more advanced peoples.[28] So, despite some scholars' views to the contrary, it is clear that Darwin's progressivist theory entailed a hierarchy of the human races. His opposition to slavery, which was deeply felt, did not mitigate his racial evaluations.[29]

Darwin's racialism never included Jews. His few scattered references to Jews contain nothing derogatory. Of some interest, though, is that he did observe that Jews and Aryans were similar in features, due, he supposed, to "the Aryan branches having largely crossed during their wide diffusion by various indigenous tribes."[30] This statement contrasts with the views of Hitler, for whom the Jews and Aryans were pure (i.e., unmixed) races—a matter discussed below. Haeckel, however, does include Jews in his hierarchical scheme.

26. Charles Darwin, *On the Origin of Species* (London: Murray, 1859), 489.

27. Charles Darwin, *The Descent of Man and Selection in Relation to Sex*, 2 vols. (London: Murray, 1871), 1:34, 239.

28. In the second edition of the *Descent*, Darwin described the extinction of the Tasmanians and the decline of the other "primitive" races of the South Pacific. See Charles Darwin, *The Descent of Man and Selection in Relation to Sex*, with an introduction by James Moore and Adrian Desmond (London: Penguin Group, [1879] 2004), 211–22.

29. Adrian Desmond and James Moore maintain that Darwin's antislavery attitude led him to postulate species descent from a common ancestor and thus establish the brotherhood of man. I am not convinced by the thesis, but even if true, this does not contradict his notion of racial hierarchy. Christian slaveholders in the American South likewise assumed common ancestry for human beings. See Adrian Desmond and James Moore, *Darwin's Sacred Cause* (New York: Houghton Mifflin Harcourt, 2009); and Robert J. Richards, "The Descent of Man: Review of *Darwin's Sacred Cause*," *American Scientist* 97 (September–October 2009): 415–17.

30. Darwin, *Descent of Man* (1871), 1:240.

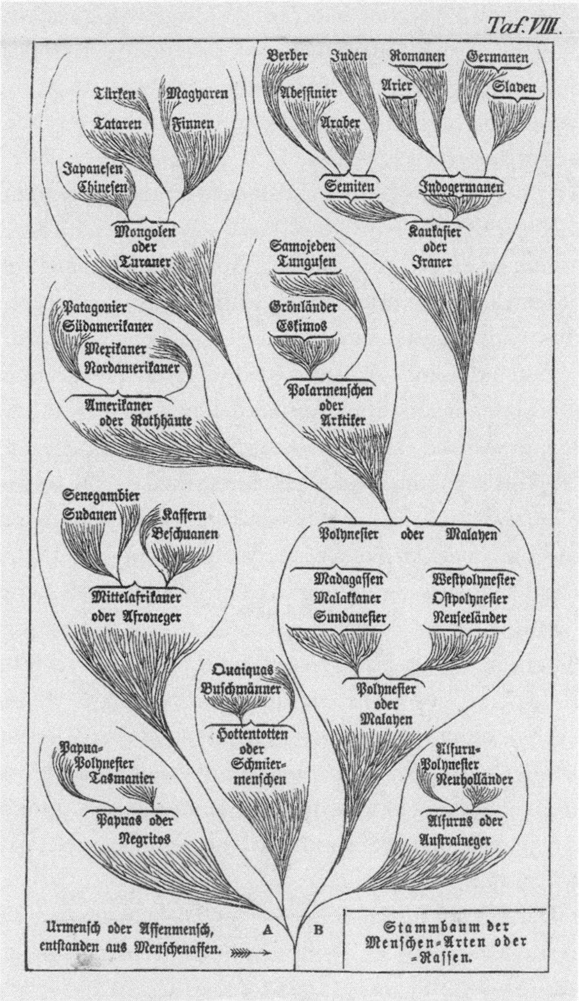


FIGURE 9.2 Stem-tree of the human species, originating in the "ape-man." From Ernst Haeckel, *Natürliche Schöpfungsgeschichte* (1868).

In the first edition of his *Natürliche Schöpfungsgeschichte* (Natural history of creation, 1868), Haeckel represented in a tree diagram nine species of human beings, along with their various races, all stemming from the *Affenmensch*, or ape-man. The vertical axis of the diagram was meant to suggest progressive development in intelligence and moral character (fig. 9.2); it showed Australians,

Hottentots, and Papuans at the lowest branches, with Caucasians occupying the highest. Not surprisingly, perhaps, the German and Mediterranean races of the Caucasian species (upper right in the diagram) are leading the other groups—except, that is, for the Berbers and the Jews, two other branches of the same species. Haeckel located the Jews at the same evolutionary level as the Germans and other Europeans—hardly the kind of judgment expected of a supposed racial anti-Semite.[31]

Haeckel spoke directly to the question of anti-Semitism. He, along with some forty other European intellectuals and artists, was interviewed in the early 1890s about the phenomenon of anti-Semitism by Hermann Bahr (1863–1934), a journalist and avant-garde playwright. Haeckel mentioned that some of his students were anti-Semitic but he explicitly disavowed that prejudice himself. He acknowledged that some nations, including Germany, were judicious in barring the immigration of Slavic Jews since they would not adopt the customs of their new countries but remained stubbornly unassimilated. He yet celebrated the *gebildeten Juden* of Germany. He is quoted by Bahr as proclaiming: "I hold these refined and noble Jews to be important elements in German culture. One should not forget that they have always stood bravely for enlightenment and freedom against the forces of reaction, inexhaustible opponents, as often as needed, against the obscurantists [Dunkelmänner]. And now in the dangers of these perilous times, when Papism again rears up mightily everywhere, we cannot do without their tried and true courage."[32] As is suggested by this quotation, Haeckel's long-term opponent was the Catholic Church, for which he had a mixture of disdain and— at least for its black-robed troops, the Jesuits—some grudging admiration.[33]

So neither Darwin nor the leading German Darwinian, Ernst Haeckel, can be accused of anti-Semitism—certainly not the kind of racism that fueled Hitler's animus and stoked the fires of the Holocaust. The belief in a racial hierarchy, assumed by both Darwin and Haeckel, also needs to be put in a larger historical context. The common presumption of higher and lower races ante-

31. Ernst Haeckel, *Die Natürliche Schöpfungsgeschichte* (Berlin: Georg Reimer, 1868), 519. In subsequent editions, Haeckel added more species and changed the location of the races in the hierarchy. In the second edition, for instance, Jews are located just a bit below the level of the Germans but still remain far ahead of most of the other races.

32. Haeckel, as quoted in Hermann Bahr, "Ernst Haeckel," in *Der Antisemitismus: Ein internationals Interview* (Berlin: S. Fischer, 1894), 62–69, quotation at 69.

33. I have explored the question of Haeckel's supposed anti-Semitism in greater detail in Richards, "Ernst Haeckel's Alleged Anti-Semitism and Contributions to Nazi Biology," *Biological Theory* 2 (Winter 2007): 97–103.

dates Darwin's work by many generations and cannot be uniquely attributed to Darwinian theory.

The pre-evolutionary naturalists Carolus Linnaeus (1707–1778), Johann Friedrich Blumenbach (1752–1840), Georges Cuvier (1769–1832), and Carl Gustav Carus (1789–1869)—all of whose works directed subsequent thought about the distinction of human races—ranked those races in a hierarchy, with Europeans, naturally, in the top position.[34] For example, Linnaeus placed the genus Homo within the order Primates (which included monkeys, bats, and sloths) and distinguished two species: *Homo sapiens* and *Homo troglodytes* (anthropoid apes). He divided *Homo sapiens* (wise man) into four varieties: American (copper-colored, choleric, regulated by custom), Asiatic (sooty, melancholic, and governed by opinions), African (black, phlegmatic, and governed by caprice), and European (fair, sanguine, and governed by laws). Linnaeus conceived such differences as expressive of divine intent.[35] Carl Gustav Carus affirmed a comparable hierarchy, though he declared that the races of mankind could not be classified with animals, as had Linnaeus. Because of their mental character, humans formed a kingdom of their own with four distinct races, each endowed with different abilities: "the people of the day" (Europeans, Caucasians, Hindus), "the people of the night" (Aethiopians—South Africans, Papuans, Australians), "the people of the eastern twilight" (Asians—Mongols and Malays), and "the people of the western twilight" (North and South American Indians).[36] The original lands of these peoples—their climate and geography—wrought effects on their anatomy, especially on skull sizes and brain formation, rendering them with different capacities for cultural attainment. The people of the day had achieved the highest development in the appreciation of beauty, truth, and goodness.[37] Although each of the groups could be located in an ascending hierarchy, human mentality remained distinctly separated from the capacities of brutes, which meant, in Carus's terms, that they

34. See, for example, Carolus Linnaeus, *Systema naturae per regna tria naturae, secundum classes, ordines, genera, species, cum characteribus, differentiis, synonymis, locis*, 3 vols. (Halle: Curt, 1760–70), 1:20–24; Johann Friedrich Blumenbach, *De generis humani varietate nativa liber*, 3rd ed. (Göttingen: Vandenhoek et Ruprecht, 1795); and Georges Cuvier, *Le régne animal*, 2nd ed., 5 vols. (Paris: Deterville Libraire, 1829–30), 1:80. I have discussed these and other hierarchical schemes in Robert J. Richards, "Race," in *Oxford Companion to the History of Modern Science*, ed. John Heilbron (Oxford: Oxford University Press, 2001), 697–98. See also Uwe Hoßfeld, *Biologie und Politik: Die Herkunft des Menschen* (Erfurt: Landeszentrale für politische Bildung Thüringen, 2011), 16.

35. Linnaeus, *Systema naturae per regna tria naturae*, 1:20–24.

36. Carus essentially reproduced the categories of Blumenbach's *De generis humani varietate nativa liber*.

37. Carl Gustav Carus, *System der Physiologie für Naturforscher und Aerzte*, 2 vols. (Dresden: Gerhard Fleischer, 1838), 1:124.

certainly did not derive from any ape forbearer, as suggested by Lamarck.[38] These racial categories of leading naturalists, established long before the appearance of Darwin's work, were mutually reinforcing of common prejudices. But the point to be made is simply that assumptions of racial hierarchy, ubiquitous in the nineteenth and early twentieth centuries, did not originate in Darwinian evolutionary theory; they were commonplaces in scientific literature since at least the eighteenth century. Darwin and Haeckel, like most other naturalists of the period, simply accepted the hierarchy and gave it an account in terms of their theoretical system.

## THE RACIAL IDEOLOGY OF GOBINEAU AND CHAMBERLAIN

At the beginning of the twentieth century, two of the most influential proponents of the theory of racial hierarchy were Joseph Arthur, comte de Gobineau (1816–1882; fig. 9.3), and Houston Stewart Chamberlain (1855–1927). Gobineau's four-volume *Essai sur l'inégalité des races humaines* (Essay on the inequality of the human races, 1853–55) was translated into several languages. It went through five German editions from 1895 to 1940 and served as the intellectual rationale for the anti-Semitic Gobineau societies that spread through Germany at the turn of the century.[39] Chamberlain's *Die Grundlagen des neunzehnten Jahrhunderts* (The foundations of the nineteenth century) flooded Germany with an amazing thirty editions from 1899 to 1944. Chamberlain was inspired by Gobineau's analysis of race and became a member of the elite Gobineau society, along with other partisans of the cult of Richard Wagner.[40] The books of Gobineau and Chamberlain helped to articulate and give form to the racial views of Hitler and his chief party philosopher manqué, Alfred

38. Ibid., 112. "Finally and chiefly it must not be thought that man has arisen from an animal (an ape, for instance, with which one sometimes classifies human beings) that has progressively developed and so has become man." Carus further refined his discussion in a work occasioned by the hundredth birthday of that great genius of the people of the day, Johann Wolfgang von Goethe: Carus, *Denkschrift zum hundertjährigen Geburtsfeste Goethe's. Ueber ungleiche Befähigung der verschiedenen Menschheitstämme für höhere geistige Entwickelung* (Leipzig: Brockhaus, 1849). Carus used the American Samuel Morton's measurement of skull sizes as one index of different intellectual capacities (19).

39. I have used the second German edition in this analysis: Joseph Arthur Grafen Gobineau, *Versuch über die Ungleichheit der Menschenracen*, trans. Ludwig Schemann, 2nd ed., 4 vols. (Stuttgart: Fr. Frommanns Verlag, 1902–4).

40. Paul Weindling provides a trenchant account of the Gobineau Society, with its elitist and nonscientific membership. See the richly nuanced Weindling, *Health, Race and German Politics between National Unification and Nazism, 1870–1945* (Cambridge: Cambridge University Press, 1989), 106–9.

FIGURE 9.3 Arthur, comte de Gobineau (1816–1882).
Lithograph from Eugen Kretzer, *Joseph Arthur, graf von Gobineau* (1902).

Rosenberg (1893–1946). For that reason, I will linger over the works of these two harbingers of the Nazi movement.

Arthur, comte de Gobineau, was born of a royalist family in 1816. His father joined the antirevolutionary forces during the Directorate and was later imprisoned by Napoleon's regime.[41] Through his early adulthood he mourned the passing of the aristocratic order and expressed in several novels, poems, and plays of the 1840s his distaste for the materialistic and crass attitudes

41. For Gobineau's family background and political orientation, I have relied on Michael Biddiss, *Father of Racist Ideology: The Social and Political Thought of Count Gobineau* (New York: Weybright and Talley, 1970).

of the rising bourgeoisie. His odd friendship with Alexis de Tocqueville (1805–1859)—with whom he had a considerable correspondence over religion, morals, and democracy—brought him into the troubled government of the Second Republic in 1849; after the coup of Louis Napoleon in 1851, he advanced to several diplomatic posts during the regime of the Second Empire (1851–71). His diplomatic work allowed him sufficient leisure time to cultivate a knowledge of Persian, Greek, and South Asian languages and civilizations, which reinforced his sentiments about a golden age of aristocratic order. He elevated his class prejudices to something quite grand: he argued that modern nations had lost the vitality characterizing ancient civilizations and that the European nations, as well as the United States, faced inevitable decline, with the French Revolution being an unmistakable sign of the end. When he learned of Darwin's evolutionary theory he disdainfully dismissed it, thinking its anemic progressivism a distortion of his own rigorously grounded empirical study; certainly the time was near, he believed, when Haeckel's phantasms of ape-men would evanesce.[42] He was assured of the decline of human societies—so palpable before his eyes during the years of political turmoil throughout Europe—and proposed a very simple formula to explain it: race mixing.

Gobineau indicated that he was moved to write his *Essai* because of the views of James Cowles Prichard (1786–1848), who argued for the essential unity of mankind and the common capacities of the various human races.[43] Gobineau wished to demonstrate, on the contrary, that while we might have to give notional assent to the biblical story of a common origin, the fundamental traits of the white, yellow, and black races were manifestly different and their various branches displayed intrinsically diverse endowments. To support this contention, he spun out, over four substantial volumes, a conjectural anthropology whose conclusions, he ceaselessly claimed, had the iron grip of natural law. The beginning of his story, he allowed, had a bit of mythical aura about it. The Adamite generation, knowledge about which trailed off into fable, begot the white race—about this the Bible seemed certain, whereas the origins of the yellow and black races went unmentioned in the sacred texts.[44] So we might assume that each of these races had independent roots, since each displayed markedly different traits.[45] The whites were the most beautiful, intelligent, orderly, and physically powerful; they were lovers of liberty and aggressively pursued it. They played the dominant role in any civilization that

42. Gobineau, *Versuch über die Ungleichheit der Menschenracen*, 1:xxxi–xxxiii
43. Ibid., xxviii–xxix.
44. Ibid., 157.
45. Ibid., 278–81.

had attained a significant culture. The yellow race was lazy and uninventive, though given to a narrow kind of utility. The black race was intense, willful, and with a dull intellect; no civilization ever arose out of the pure black race. Each of the three races had branches with somewhat different characters. So, for instance, the white race comprised the Assyrian, Celtic, Iberian, Semitic, and Aryan stocks. These stocks had intermingled to produce the great civilizations of the past—Gobineau discriminated some ten such ancient civilizations.[46] The Greek civilization, for example, arose from the Aryan stock with a tincture of the Semitic. High attainment in culture, science, and the arts had only existed, however, where there was a large admixture of the Aryan. Even the Chinese, in his estimation, derived from an Aryan colony from India. Had these branches of the white race remained pure, their various ancient civilizations would still be flourishing. But racial mixing caused an inevitable degradation of their character.

Gobineau postulated two contrary forces operative on the races of mankind: revulsion for race mixing, especially powerful among the black groups, and a contrary impulse toward intermarriage, which oddly was characteristic of those peoples capable of great development.[47] As a result of the impulse to mate with conquered peoples, the pure strains of the higher stocks had become alloyed with the other strains, the white race being constantly diluted with the blood of inferior peoples, while the latter enjoyed a boost from white blood. Contemporary societies, according to Gobineau, might have more or less strong remnants of the hereditary traits of their forbearers, but they were increasingly washed over as the streams of humanity ebbed and flowed. The modern European nations thus lost their purity, especially as the white component had been sullied in the byways of congress with the yellow and black races. So even the modern Germans, who still retained the greatest measure of Aryan blood and yet carried the fire of modern culture and science—even the Germans had begun to decline and would continue to do so as the tributaries of hybrid stocks increasingly muddied the swifter currents of pure blood.

Despite Gobineau's theories of race and his influence in Germany, he was no egregious anti-Semite, at least not of the sort that so readily adopted his views. He regarded the Jews as a branch of the Semites, the latter being a white group that originally extended from the Caucasus Mountains down through the lands of the Assyrians to the Phoenician coast. The Hebrews, as he preferred to call the Jews, retained their racial purity up to the time of the reign of

46. Ibid., 287–90.
47. Ibid., 38.

King David, a period when so many other, less worthy, peoples were brought into the kingdom: "The mixing thus pressed through all the pores of Israel's limbs." As a consequence, "the Jews were marred through mating with blacks, as well as with the Hamites and Semites in whose midst they were living."[48] In short, the Jews fared no better and no worse than other groups of originally pure stock; like them, the Jews enjoyed for a while the advantages of a homogeneous population and then slipped silently down the racial slope into their current mongrel state.[49]

The theme of cultural degradation due to race mixing echoed through the decades after the publication of Gobineau's treatise. Richard Wagner, who became a friend and correspondent of Gobineau, anticipated the dangers of racial decline, though, like the poet Friedrich Schiller (1759–1805), believed that art might reverse the decline, at least for the German people. Americans also heard the unhappy knell. Madison Grant (1865–1937), a New York lawyer, with biological and anthropological acumen on a level below that even of his French predecessor, pressed the same concerns in a comparably conjectural study, *The Passing of the Great Race* (1916), the German edition of which was found in Hitler's library.[50] Grant thought the superior Nordic race—the true descendant of the Aryan peoples—to be endangered by crossbreeding. He thought the proximate danger to Aryan purity came from the two lower stocks of the Caucasian group—the Alpine race (eastern Europeans and Slavs) and the Mediterranean race (stemming from the southern areas of Asia minor and along the coasts of the inland sea), thus the swarthy Poles, Czechs, and Russians and the even more swarthy Spaniards, Italians, and Greeks. Unmistakable signs indicated the decline of the American civilization: simplified spelling and incorrect grammar told the story, for Grant, of decay from Nordic standards.[51] Even more alarming were the Polish Jews swarming into New York City—the *cloaca gentium*, in terms borrowed from Chamberlain: the Jews

48. Ibid., 2:92–93.

49. By contrast, his German translator and biographer Ludwig Schemann, in *Von deutscher Zukunft* (1920), turned Gobineau's thesis of the dangers of racial decline against the Jews. Schemann detected in the Jews "a lethal danger for our material life as well as for our spiritual and ethical life." The Jews, he contended, "should be regarded as an alien people in our civic life." As quoted in Hoßfeld, *Biologie und Politik*, 38.

50. Madison Grant, *The Passing of the Great Race or the Racial Basis of European History* (New York: Charles Scribner's Sons, 1916). Hitler's library contained the German translation, *Der Untergang der großen Rasse*, trans. Rudolf Polland (Munich: Lebmanns Verlag, 1925). See Timothy Ryback, *Hitler's Private Library* (New York: Vintage Books, 2010), 97. Since Hitler's copy does not contain any markings, and he doesn't mention Grant by name, it's uncertain whether he actually read the book. Further, the first volume of *Mein Kampf* was finished in early 1925, and the translation of Grant's work came out in summer 1925.

51. Grant, *Passing of the Great Race*, 6.

wore the Nordic's clothes and stole his women, thus genetically obliterating his commanding stature, blue eyes, blond hair, and Teutonic moral bearing.[52] (There appears to be no accounting for Nordic women's taste in men.) The German nation fared little better; through miscegenation it had suffered a large decline in the number of pure Teutons.[53] Grant played in syncopated harmony the American version of Gobineau's tune. But the most influential orchestrator of this theme at the turn of the century, done in Wagnerian style, was Houston Stewart Chamberlain.

Chamberlain, born in 1855, descended from the lesser British aristocracy and from money on both sides of his family.[54] His father, mostly absent from his life, fought in the Crimean War, serving as an admiral of the British fleet. After his mother suddenly died, he and his two brothers were shipped off to Versailles to live with a grandmother and aunt. In 1866, to reintroduce him to his native heritage, his father enrolled the ten-year-old, French-speaking lad in an English school, but ill health kept him there only a few years. The boy returned to France, where his schooling was taken over by a German tutor, who instilled a love of the language and culture of Germany. After three years his tutor took up a post back in his native land; Chamberlain, now thirteen, saw to his own education, reading promiscuously in the literature of Germany, France, and England and cultivating an interest in the solitary science of botany. His father died in 1878, leaving him with a decent income and freedom to marry a woman whom he had met when a teenager of sixteen and she twenty-six. The nuptials occurred three years later. He now worried about a formal education. His self-tutelage was sufficient to win him a place in the natural science faculty at the University of Geneva, from which he graduated with distinction in 1881. While at Geneva he came under the autocratic sway of Karl Vogt (1817–1895), whom he thought too influenced by the experience of the revolutions of 1848. Vogt was an evolutionist, although according to Chamberlain's reckoning, he was mistrustful of Darwinism and Haeckelianism.[55] The young student pursued a doctoral thesis in plant physiology at Geneva but interrupted his study after two years due to a free-floating nervous indisposition. His attempt at a stock brokerage business met quiet failure. With the aid of additional funds from

52. Ibid., 81. The expressive phrase *cloaca gentium*—sewer of the races—appears to have come from Chamberlain, who used it to refer to Rome. See Houston Stewart Chamberlain, *Die Grundlagen des neunzehnten Jahrhunderts*, 2 vols. (Munich: Bruckmann, 1899), 1:286.

53. Grant, *Passing of the Great Race*, 166.

54. For the details of Chamberlain's life, I have relied on the fine biography by Geoffrey Field, *Evangelist of Race: The Germanic Vision of Houston Stewart Chamberlain* (New York: Columbia University Press, 1981).

55. Houston Stewart Chamberlain, *Lebenswege meines Denkens* (Munich: Bruckmann, 1919), 93.

his aunt, he continued private study, especially in German philosophy and literature; Kant and Goethe became his loadstars. Then he discovered Richard Wagner, and his glittering firmament was fixed.

Shortly after he was married in 1878, Chamberlain and his wife Anna attended the premier of *Der Ring des Nibelungen* in Munich, an event that ignited what would become an ever-growing passion for the numinous music and deranged doctrines of the great composer. In 1882, the couple visited the consecrated ground of Bayreuth, where they heard *Parsifal* three times. He wrote his aunt that the "overwhelming beauty" simply stunned him (*mich einfach verstummen machte*).[56] Not only did the aesthetic power of the music transfix him, but his fervent Christianity became alloyed with the mystical theology fueling the legends of questing knights and battling gods. He enrolled as a member of the Wagner Society (Wagner-Verein), formed after Wagner's death in 1882, and helped found a new French journal devoted to the art of the composer. His many articles for the journal drew him closer to Cosima Wagner (fig. 9.4), second wife of the maestro, daughter of Franz Liszt, and titular head of the inner circle of the cult, which fed on the racial theories of Gobineau, now growing into Teutonic glorification and pernicious anti-Semitism. The measure of Chamberlain's devotion not simply to the music but to the mystical association of Wagner with the German spirit can be taken by the extent of his labors: he wrote four books and dozens of articles on the man and his music during the short period between 1892 and 1900.[57] The more significant measure, perhaps, was the kindling of his admiration for, if not burning love of, Wagner's youngest daughter, Eva, whom he married in 1908 following an expensive divorce from his first wife.

After moving from Dresden to Vienna in 1889—and still relying on the financial kindness of his aunt—Chamberlain renewed his intention to finish a doctorate in plant physiology. He started attending lectures at the university, especially those of the botanist Julius Wiesner (1838–1916), with whom he became friendly, despite Wiesner's Jewish ancestry. With the encouragement of Wiesner, he resurrected extensive measurement experiments on the movement of fluids in plants that he had originally conducted in Geneva. Since his ner-

56. Chamberlain to Harriett Chamberlain (31 July 1882), in *Houston Stewart Chamberlain Briefe, 1882–1924, und Briefwechsel mit Kaiser Wilhelm II*, ed. Paul Pretzsch, 2 vols. (Munich: Bruckmann, 1928), 1:1.

57. Chamberlain's books on Wagner are *Das Drama Richard Wagner's. Eine Anregung* (Vienna: Breitkopf und Härtel, 1892); *Richard Wagner. Echte Briefe an Ferdinand Praeger* (Bayreuth: Grau'sche Buchhandlung, 1894); *Richard Wagner* (Munich: F. Bruckmann,1896); and *Parsifal-Märchen* (Munich: F. Bruckmann, 1900). Each of these went through multiple editions and translations.

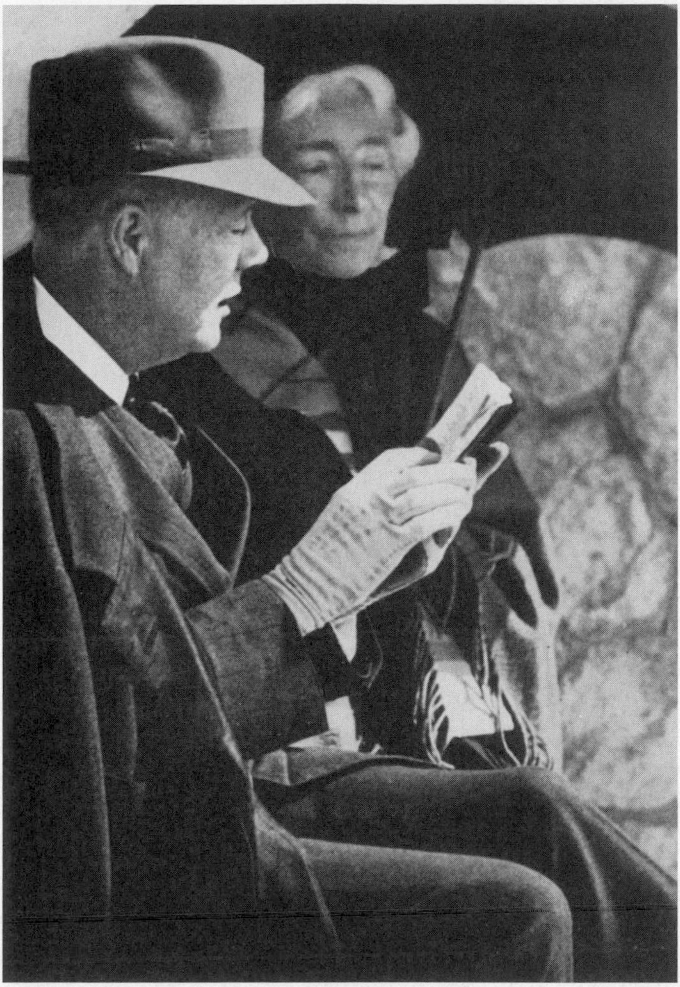

FIGURE 9.4 Cosima Wagner (1837–1930) and her son-in-law, Houston Stewart Chamberlain (1855–1927). Photo, 1913, from *Cosima Wagner und Houston Stewart Chamberlain, Briefwechsel.*

vous condition precluded further experimental work, he now put his original findings into a broad historical and philosophical context, arguing that no adequate mechanistic account could be given of the rise of sap in plants and its resistance to falling back.[58] We must assume, he contended, that vital forces

58. Chamberlain, *Lebenswege meines Denkens*, 119–20.

are at work. Whether these forces operated extrinsic to the molecular structure or were internal to it, the evidence confirmed their presence: mechanical forces alone could not lift the sap in trees the 150 or 200 feet of their height.[59] Despite an insatiable mania for publishing (his *Schreibdämon*, as he called it), the writing of the dissertation was desultory; the work finally appeared in 1897, although it was not submitted for a degree. Immediately on its publication, Chamberlain began the composition of his masterwork, *Die Grundlagen des neunzehnten Jahrhunderts* (later published in English as *The Foundations of the Nineteenth Century*, and referred to hereafter as *Foundations*), which would eventually flood Germany with a rich farrago of Goethean sentiment, Kantian epistemology, Wagnerian mysticism, and Aryan anti-Semitism. The medley echoed through the German reading public for almost half a century.

While Gobineau maintained that the races originally were pure but tended to degenerate over time because of miscegenation, Chamberlain contended that purity of race was achieved over long periods of time; once achieved, however, it could be endangered by race mixing.[60] His notion of race was quite loose, insofar as the Greeks, Romans, Iranians, Chinese, English, French, Jews, Aryans (or Germans) all formed, in his estimation, distinct races. His test of race was the direct, intuitive experience of the other, rather than any craniometric measures. He was vague about the origins of human beings, simply observing that as far as history testified, human beings have always existed.[61] He dismissed as a "pseudo-scientific fantasy" Haeckel's argument that the human races descended from apelike forbearers.[62]

For Chamberlain, the two principal races that achieved purity and retained it were the Aryan and the Jewish. The Aryans, which in their more recent incarnation he referred to as Germans, were the bearers of culture, science, and the arts. Their mental accomplishments flowed from blood, he argued (or really, simply stipulated). In a wonderful piece of quasi-idealistic morphology, he described the real German as having an ideal type: "great, heavenly radiant eyes, golden hair, the body of a giant, harmonious musculature, a long skull [and]... high countenance."[63] All of this notwithstanding, individual Germans might be dark-haired, brown-eyed, and small of stature. (One had to see the blond giant standing behind the form, for example, of the puny chicken-farmer

59. Houston Stewart Chamberlain, *Recherches sur la sève ascendante* (Neuchâtel: Attinger Frères, 1897), 6–8.

60. Chamberlain, *Grundlagen*, 1:266–67.

61. Ibid., 277.

62. Ibid., 122n.

63. Ibid., 496.

with dark, receding hair—Heinrich Himmler.) Against the blond giant stood the threatening Jew. Chamberlain devoted 135 continuous pages to dissecting the Jewish type, its physiology and character. So distinct were the racial traits that one could be certain that Christ was not a Jew, a view that Hitler took over from Chamberlain.[64] Throughout the *Foundations*, this Anglo-German would vacillate between referring to the Jews as a pure race, meaning relatively permanent, but also of a "mongrel character [Bastardcharakter]."[65] That character displayed the typical attitudes his fellows had come to associate with Jews: materialistic, legalistic, limited in imagination, intolerant, fanatical, and with a tendency toward utopian economic schemes, as found, for instance, in Marxism.[66] The Jews' very "existence is a sin [Sünde]; their existence is a transgression against the holy laws of life."[67] Thus any mating between Jew and Aryan could only corrupt the nobility of the latter: the Jewish character "is much too foreign, firm, and strong to be refreshed and ennobled by German blood."[68] This could only mean a struggle between the Aryans and the Jews, "a struggle of life and death [ein Kampf auf Leben und Tod]."[69]

Chamberlain used the trope of racial struggle frequently in the *Foundations.* Indeed, the phrase usually identified with Darwinian theory, "struggle for existence" (*Kampf ums Dasein*), appears eight times in the *Foundations.* The single word "struggle" (*Kampf*) turns up 112 times. But these terms were not markers of Darwin's theory of natural selection. Chamberlain rejected Darwin's conception completely, comparing it to the old, discredited "phlogiston theory."[70] Not only did he dismiss Darwin's main explanatory device, but he rejected transmutation of species altogether. After all, it was an idea already refuted in advance by Kant.[71] Darwin's theory, however, continued perniciously to affect all it touched. Chamberlain wrote Cosima Wagner at the time of the composition of the *Foundations*: "this hair-raising absurdity poisons not only natural science but the whole of human thought: Darwinism rules everywhere,

64. Chamberlain goes through some conceptual contortions to reach this conclusion. See ibid., 217–20. Hitler adopted the same theory, namely, that "Christ was certainly not a Jew, but a Galilean of Aryan descent." See Adolf Hitler, *Monologe im Führer-Hauptquartier, 1941–1944*, ed. Werner Jochmann (Munich: Albrecht Knaus, 1980), 96 (21 October 1941). This latter volume recovers Hitler's "Table Talk," stenographic transcripts ordered by Martin Bormann of the leader's conversations.

65. Chamberlain, *Grundlagen*, 1:372.

66. Ibid., 415.

67. Ibid., 374.

68. Ibid., 325.

69. Ibid., 531.

70. Ibid., 2:805.

71. Ibid., 1:25.

corrupting history and religion; it leads to social idiocy; it degrades judgment about men and things."[72]

In a letter of advice to a young student, Chamberlain contended that while some of Darwin's observations might be empirically helpful, his theory "is simply poetry [einfach eine Dichtung]; it is unproven and unprovable." Anyone with the least tincture of metaphysics would understand the impossibility of solving the world puzzles by evolution.[73] The main difficulty—as he detailed in manuscripts composed at the time of the *Foundations*—has to do with the integrity of form. Taking his cue from Georges Cuvier, Goethe, and Kant, Chamberlain argued that our direct, intuitive experience revealed only two archetypal forms in the plant world and eight in the animal world (e.g., radiate animals, articulate animals, vertebrate animals, etc.) governed by laws of formation (*Bildungsgesetze*). These fundamental forms simply could not pass into one another, otherwise we would have the ape being a cousin of the tree it was climbing. Moreover, animal forms exhibited an integral correlation of their constituent parts, constrained within certain limits of variability, such that any radical change of a part would collapse the harmony of the whole, and radical changes in an animal's form would fatally disrupt its relation to other animals. Thus transmutation of forms, as Lamarck, Darwin, or Weismann conceived it, would be impossible.[74] Chamberlain's racism and conception of struggle of races owed no theoretical debt to Darwin, Haeckel, Weismann, or any other of the Darwinians but rather chiefly to Gobineau, Kant, Goethe, and Wagner—insofar as responsibility might be thought transitive.[75]

## CHAMBERLAIN AND HITLER

Hitler's racial infections derived from many sources—particularly the seething political pool he threw himself into while in Vienna as a young, aspiring art student and feckless vagabond. But in *Mein Kampf*, no placid reservoir of ideas, he seems to have deployed slightly less agitated concepts to structure his

72. Chamberlain to Cosima Wagner (9 March 1896), in *Cosima Wagner und Houston Stewart Chamberlain im Briefwechsel, 1888–1908*, ed. Paul Pretzsch (Leipzig: Philipp Reclam, 1934), 478.

73. Chamberlain to Karl Horst (31 October 1895), in *Chamberlain Briefe*, 1:26–27. The phrase "world puzzles" was obviously an oblique reference to Haeckel's book *Welträtsel*.

74. These are the conclusions Chamberlain drew in two manuscripts from the years 1896 and 1900. They were published by his friend Jakob von Uexküll shortly after his death. See Houston Stewart Chamberlain, *Natur und Leben*, ed. J. von Uexküll (Munich: Bruckmann, 1928), 102–68.

75. In matters of morphology, Chamberlain said his masters were Goethe and Kant. See Chamberlain, *Lebenswege meines Denkens*, 122.

considerations of race. His promiscuous mind culled these ideas from many quarters, but one in particular stands out—those theories and conceptions of Houston Stewart Chamberlain—and not by accident.

Hitler likely first encountered Chamberlain's *Foundations* sometime between 1919 and 1921, when he read the work at the National Socialist Institute Library in Munich.[76] He met the man himself shortly thereafter in Bayreuth. Chamberlain moved to Bayreuth after his marriage to Eva Wagner in 1909, and there he served to help reorganize the finances of the Festspiele and edit the *Bayreuther Blätter*, which carried articles on the art of the master interlaced with observations on the perfidy of Jews. As the leader of the growing German Workers Party, Hitler traveled to Bayreuth in late September 1923 to attend a political rally. While in the city, he was invited by the Wagner family to visit and worship at Wahnfried, the maestro's home and shrine. Chamberlain spoke extensively with the man over two days and was so impressed that he wrote the lederhosed politician an amazingly fulsome letter, which Hitler never forgot. The long letter of 7 October read in part:

> You are certainly not as you have been described to me, namely as a fanatic [Fanatiker]; rather I would call you the very opposite of a fanatic. A fanatic overheats the head, while you warm the heart. The fanatic wishes to smother you in words; you want to convince, only convince. . . . My faith in Germanness [Deutschtum] has never wavered for a moment. But my hopes—I will confess—had ebbed. With one blow, you have transformed the core of my soul. That Germany in the hour of her greatest need has given birth to a Hitler, that shows her vital essence.[77]

On the occasion of Hitler's thirty-fifth birthday, celebrated the next year in prison (fig. 9.5), Chamberlain published an open letter in which he extolled this man, so different from other politicians, a man who "loves his German people with a burning passion." "In this feeling," he professed, "we have the central point of his whole politics, his economics, his opposition to the Jews, his battle against the corruption of values, etc."[78] After his release from jail, Hitler visited Chamberlain on several occasions and mourned him at his

76. Ryback, *Hitler's Private Library*, 50.

77. Chamberlain to Adolf Hitler (7 October 1923), in *Chamberlain Briefe*, 2:124–25.

78. The letter was originally published in *Deutsche Presse*, nos. 65–66 (20–21 April 1924), 1; reprinted in *Houston Stewart Chamberlain, Auswahl aus seinen Werken*, ed. Hardy Schmidt (Breslau: Ferdinand Hirt, 1935), 66.

FIGURE 9.5 Adolf Hitler (1889–1945) in Landsberg Prison, 1924. (Courtesy of Getty Images)

funeral.[79] In the depths of World War II, Hitler recalled with extreme gratitude visiting Bayreuth for the first time and meeting Chamberlain. In his "Table Talk"—conversations ordered by Martin Bormann to be stenographically recorded—Hitler mentioned that "Chamberlain's letter came while I was in

79. Hitler visited Chamberlain several more times in Bayreuth, in spring and summer 1925 and again in November and May 1926, when the old man was in very poor health. Chamberlain died on 9 January 1927. Hitler, representing the Workers Party, attended the funeral services.

jail. I was on familiar terms with them [Chamberlain and the Wagner family]; I love these people and Wahnfried."[80] It was while in jail, comforted as he was by Chamberlain's recognition, that he composed the first volume of *Mein Kampf*.

## *Mein Kampf*

In early November 1923, Hitler, leading the German Workers Party and its quasi-military wing, the Sturmabteilung or SA, attempted to overthrow the Munich municipal government, hoping thereby to galvanize the masses and march on Berlin. This Beer Hall Putsch, as it was called, failed miserably, and the following spring, Hitler and his deputy Rudolf Hess, along with other conspirators, were sentenced to five years in jail. Because of sympathy for Hitler's effort to "save the nation," he and Hess were confined to a minimum security compound, Landsberg Prison. During his stay, Hitler was allowed unlimited visitors, any number of books, and his faithful dog. He famously called this time in jail his "higher education at state expense."[81] While in jail he was visited often by Alfred Rosenberg, who had become party chairman in the leader's absence. Rosenberg at this time was completing his *Myth of the Twentieth Century*, a book he regarded as a sequel to Chamberlain's *Foundations of the Nineteenth Century* and that Hermann Goering (1893–1946) regarded as a "philosophical belch."[82] Presumably Rosenberg and Hitler spoke of mutual concerns, since both were authoring books with similar political and racial themes. Hitler began the composition of *Mein Kampf* in July 1924, and it quickly became inflated into two large volumes by the next year. He initially wanted to title it *A Four and a Half Year Battle against Lies, Stupidity and Cowardice*, but finally shortened the title simply to *My Battle—Mein Kampf*. The book brewed up a mélange of autobiographical sketches, a theory of race, a declaration of the need to expand the land of the Germans, principally to the east, and foreign policy exhortations to restore the honor and power of the nation. Flavoring the stew throughout was the bitter vitriol of scorn for those who had destroyed the means to win the last war and connived to push the nation into collapse after the war—the Jews, capitalists,

80. Hitler, *Monologe im Führer-Hauptquartier*, 224. It's unclear which of the two letters Hitler is referring to—the personal letter or the open letter published while he was in Landsberg Prison.

81. As quoted in Ryback, *Hitler's Private Library*, 67. The some 12,000 volumes of Hitler's libraries, recovered by American forces after the war, now reside in the Library of Congress; eighty others are in the Brown University library, souvenirs of a returning soldier.

82. Alfred Rosenberg, *Der Mythus des 20. Jahrhunderts* (Munich: Hoheneichen Verlag, 1930). The remark by Goering is quoted in Richard Evans, *The Third Reich in Power* (New York: Penguin Books, 2005), 138.

and Bolsheviks. The first volume of *Mein Kampf* appeared in summer of 1925, sometime after Hitler's parole the previous December; he had served only about seven months of his sentence. The second volume was finished in 1925 and published the next year.[83]

Quite a few conservative critics, whom I've cited at the beginning of this chapter, have contended that Hitler's *Mein Kampf* expresses a racial theory that virtually comes straight from the pages of Darwin's *Origin of Species*—or at least from those pages as reauthored by Ernst Haeckel. Yet neither Darwin's nor Haeckel's name appears in Hitler's book—quite surprising if the debt to these individuals is supposed to be profound. Indeed, the only name carrying any scientific weight that Hitler cites in *Mein Kamp* is that of Houston Stuart Chamberlain, his supporter and an avowed anti-Darwinian.[84] Perhaps the debt is silent, but nowhere does Hitler even use the terms *Evolutionslehre*, *Abstammungslehre*, *Deszendenz-Theorie*, or any word that obviously refers to evolutionary theory. If Hitler's racial views stemmed from Darwinian theory, without perhaps naming it, one would at least expect some term in general use for evolutionary theory to be found in the book—but not so. Admittedly, if you read Weikart's two books—*From Darwin to Hitler* and *Hitler's Ethic*—you will find several passages, translated from Hitler's German, that use the word "evolution." Also, Weikart relentlessly refers to Hitler as an evolutionist. Weikart, however, has played a sly trick. He generally translates the common German term *Entwicklung* as "evolution," though the usual meaning and ordinary translation would be "development." The term had been used for "evolution" in earlier German literature, just as "development" had been similarly employed in English literature. *Entwicklung* had been commonly used in biological literature to refer to embryological development. By the end of the nineteenth century, the terms *Entwicklung* and "development" as referring to species' evolution had declined in use in both Germany and England, though in German *Entwicklungslehre* would still be used to mean the theory of evolution; that compound, however, never appears in Hitler's book. In *Mein Kampf*, Hitler used *Entwicklung* in ways that make it obvious he did not mean biological evolution, for example, when he talked about "industrial development" (*industrielle Entwicklung*).[85] There are only two instances—though not in *Mein Kampf*—in which Hitler clearly mentions the theory of evolution. I will consider those instances below.

83. I have used the 1943 edition of *Mein Kampf*, which prints both volumes of the book as one: Adolf Hitler, *Mein Kampf* (Munich: Verlag Franz Eher Nachf., 1943).

84. Ibid., 296.

85. Ibid., 156.

Perhaps, however, Hitler's racial theory was yet indebted to Darwin's ideas, but without any verbal signposts. In the first section of this chapter, I indicated three essential features of Darwin's theory that anyone adopting the theory would necessarily embrace: (1) that the races are hierarchically ordered; (2) that species have descended from earlier species with modification; and (3) that such transmutation was, for the most part, under the aegis of natural selection. When Weikart, Berlinski, and many others read Hitler's book, they claim that Darwinian ideas leap out at them. But just what are those ideas? Though both Hitler and Darwin believed in a hierarchy of races, that's hardly a reliable indicator that the German leader embraced concepts of evolutionary biology: as I indicated in the second section, naturalists from Linnaeus in the mid-eighteenth century to individuals like Gobineau in the mid-nineteenth—all writing before Darwin's *Origin*—adopted hierarchical schemes as part of their scientific purview—and, of course, popular prejudice made racial scaling ubiquitous. More proximately, assumptions of racial hierarchy structured Chamberlain's conceptions—conceptions that owed no debt to Darwinism; these conceptions clearly made their impact on Hitler. Thus there were a myriad of sources of a non-Darwinian or anti-Darwinian character that might have stimulated Hitler to formalize his ideas of racial hierarchy. But if we go to the heart of the matter—the descent of species over time—we find nothing in *Mein Kampf* that remotely resembles such a notion. Quite the contrary. But before exploring that contrary evidence in *Mein Kampf*, let us consider evidence from outside the book.

In Hitler's "Table Talk," the German leader was recorded as positively rejecting any notion of the descent of human beings from lower animals. In the late evening of 25–26 January 1942, he remarked that he had read a book about human origins and that he used to think a lot about the question. He was particularly impressed that the ancient Greeks and Egyptians cultivated ideas of beauty comparable to our own, which could not have been the case were these peoples quite different from us. He asked: "Whence have we the right to believe that man was not from the very beginning [Uranfängen] what he is today? A glance at nature informs us that in the realm of plants and animals alterations and further formation occur, but nothing indicates that development [Entwicklung] within a species [Gattung] has occurred of a considerable leap of the sort that man would have to have made to transform him from an apelike condition to his present state."[86] Could any statement be more explicit? Hitler

86. Hitler, *Monologe im Führer-Hauptquartier*, 232 (25–26 January 1942). Hitler's German is an inelegant tangle, even granted that "Table Talk" records spontaneous conversations. Here's the original:

simply rejected the cardinal feature of Darwin's theory as applied to human beings. How could Darwin's conception have been responsible for Hitler's racial theory regarding human beings when that conception was in fact completely rejected by the latter?

It is not certain to what book on human origins Hitler might have been referring in the conversation during that late January evening. But after his rejection of descent theory, he immediately discussed the "world-ice theory" (*Welteislehre*) of Hanns Hörbiger (1860–1931). Hörbiger was an engineer and amateur astronomer who, in his book *Glazial-Kosmogonie* (1913), concocted a theory—which came to him in a vision—whereby an icy, dead star fell into a larger one, resulting in the creation of several planetary systems, of which ours was one. The earth, so the theory went, had a number of icy moons that periodically crashed into it, causing a series of catastrophes. About ten thousand years ago, another moon spiraled into the earth, causing the last global ice age.[87] As these ideas were elaborated by other catastrophists, they included beliefs that an original Aryan civilization existed before ours and that after the impact of that last icy moon, the saved remnants retreated to the high plateaus of Tibet. When things warmed up, these individuals came down from the mountains and eventually reestablished culture. SS chief Heinrich Himmler (1900–1945) even sent a research team to Tibet to recover the remains of that Aryan civilization.[88] Karl Rode (1901–1944), professor of geology and paleontology at Breslau, urged that world-ice theory was not merely a cosmological hypothesis but an Ur-Germanic "worldview" (*Welt-Anschauung*) complementary to that of National Socialism.[89] Hitler, for his part, contended that world-ice theory was the only assumption that made sense of the sophistication of

---

"Woher nehmen wir das Recht, zu glauben, der Mensch sei nicht von Uranfängen das gewesen, was er heut' ist? Der Blick in die Natur lehrt uns, daß im Bereich der Pflanzen und Tiere Veränderungen und Weiterbildungen vorkommen, aber nigrends zeigt sich innerhalb einer Gattung eine Entwicklung von der Weite des Sprunges, den der Mensch gemacht haben müßte, sollte er sich aus einem affenartigen Zustand zu dem, was er ist, fortgebildet haben!"

87. Hanns Hörbiger and Philipp Fauth, *Glazial-Kosmogonie* (Leipzig: R. Voigtländers Verlag, 1913).

88. See Christopher Hale, *Himmler's Crusade: The Nazi Expedition to Find the Origins of the Aryan Race* (New York: Wiley, 2003), 117–19.

89. Karl Rode, "Welt = Anschauung!" *Zeitschrift für die gesamte Naturwissenschaft* 2 (1936–37): 222–31. See also Christina Wessely, "Welteis. Die 'Astronomie des Unsichtbaren' um 1900," in *Pseudowissenschaft. Konzepte von Nicht/Wissenschaftlichkeit in der Wissenschaftsgeschichte*, ed. D. Rupnow, V. Lipphardt, J. Thiel, and C. Wessely (Frankfurt am Main: Suhrkamp, 2008), 155–85. Wessely shows that although Höbiger had little success convincing the leading astronomers and geologists of his theory after the First World War, yet several popular societies (Welteis-Vereine) in Germany and Austria spread the word through evening lectures and an enormous number of books. Newspapers and illustrated magazines also informed a curious public. She observes that Heinrich Himmler in particular lent the theory support.

Greek and Egyptian civilizations, and he even planned a museum that would celebrate Hörbiger, along with Ptolemy and Kepler.[90] While the world-ice theory, with its multitude of catastrophes, made sense to the German leader, it certainly would not have made sense to Darwin or Haeckel, who proposed gradualistic changes in the earth's geology and organic life such that human beings progressively evolved from apelike predecessors and slowly achieved greater intelligence and more elaborate culture. Clearly, Hitler simply rejected an essential component of Darwinian theory.

But wait a while. Weikart insists that the quoted passage from Hitler's "Table Talk" is uncharacteristic. He cites instead a passage from Hitler's speech in 1933 at Nuremberg, in which Hitler asserted: "The gulf between the lowest creature which can still be styled man and our highest races is greater than that between the lowest type of man and the highest ape." Weikart proposes that Hitler had thus essentially erased the "biblical distinction between man and other creatures."[91] Weikart suggests that this lonely remark from Nuremberg, with its supposed eradication of the distinction between man and beast, indicates the German leader's acceptance of evolution. Well, not quite. That Hitler thought the races formed a hierarchy is hardly news; it carries no suggestion of a belief in transmutation, as I have already indicated. Moreover, any slaveholding Christian in the American South could have made an observation similar to Hitler's. They clearly held black slaves to be exceedingly low in the divine hierarchy, yet still human beings. Hitler's remark seems a paraphrase of the anti-Darwinian Gobineau, who had repeated the common prejudice: "The black variety [i.e., race of human beings] is the lowest and stands on the bottom rung of the ladder. The character of an animal, which is impressed on the form of their pelvis, distinguishes them from the moment of birth to their maturity. Mentally they never move beyond the narrowest circle."[92] Though Gobineau likened the black race to lower animals, he regarded them nonetheless as human beings; Gobineau, as I've indicated, completely rejected Haeckel's ape-man hypothesis. Hitler's differential evaluation of the races hardly eliminates the distinction between human beings and lower animals.

The only other time, at least that I'm aware, in which Hitler clearly refers to evolution is in his "Table Talk" of October 1941, when he excoriated the Church for what he took as its opposition to science. He noted that the schools

90. Hitler, *Monologe im Führer-Hauptquartier* (25–26 January 1942), 232.
91. Weikart, *Hitler's Ethic*, 47.
92. Gobineau, *Versuch über die Ungleichheit der Menschenracen*, 1:278.

allowed the absurdity of having religious instruction in which biblical creation was taught during one class, and then, in the next, a natural science lesson would substitute the theory of evolution (*Entwicklungstheorie vertreteten wird*) for the Mosaic story. Hitler added that as a child he was confronted with similar contradictions between science and religion. He contended that while it was not incorrect to regard God as creator of the lightning bolt, one should not take such a notion literally; rather, it would be more profoundly pious (*tiefinnerlich fromm sein*) to find God in everything (*im Gesamten*).[93] That Hitler was aware of evolutionary theory, of course, is true—after all, he explicitly rejected human evolution some weeks later in January 1942. The racial worries saturating *Mein Kampf* have nothing to do with transmutation of species but rather its opposite.

Hitler's overriding racial concern in *Mein Kampf* was purity. He maintained that a general drive toward racial homogeneity, toward "racial purity" (*allgemein gültigen Triebes zur Rassenreinheit*), characterized all living organisms. This drive was exemplified by the uniformity and stability of species: "The consequence of this racial purity [Rassenreinheit], which is characteristic of all animals in nature, is not only a sharp separation of the particular races externally, but also in their uniformity of the essence of the very type itself. The fox is always a fox, the goose a goose, the tiger a tiger, and so on."[94] Of course for a Darwinian, there is no "essence of the very type"; the fox was not always a fox, the goose not always a goose, and in future they would not remain fixed in their types. Fixity of type is the very antithesis of a theory that contends species are not static but vary and are transformed into other species over time. Darwin's principle of diversity, which he regarded as important as natural selection, maintains that there is a general tendency of varieties and species to diversify, that is, to become heterogeneous as opposed to maintaining homogeneity.[95] Weikart's claim that Hitler "believed that humans were subject to immutable evolutionary laws" simply cannot be true.[96]

Racial purity became endangered by race mixing, especially the sullying of the higher Aryan type with the lower Jewish. Reflecting the warnings of Gobineau and Chamberlain, Hitler specified the extreme danger of miscegenation for the race of higher culture: "Historical experience offers numerous examples. It shows in awful clarity that with every mingling of blood of Aryans with lower peoples, the resulting consequence is the end of the culture bear-

93. Hitler, *Monologe im Führer-Hauptquartier* (24 October 1941), 103.
94. Hitler, *Mein Kampf*, 312.
95. See chapter 3 in the present volume.
96. Weikart, *Hitler's Ethic*, 3.

ers."[97] Hitler was assured that "all great cultures of the past were destroyed because the original, creative race died off through blood poisoning"[98]—the diagnosis of Gobineau and Chamberlain. This aspect of Hitler's argument needs to be emphasized. The Aryans, Hitler maintained, were the original bearers of culture—another verse of the gospel according to Gobineau and Chamberlain—and they propagated art and science to the rest of the world. The pure blood of the Aryans could not be improved upon, only degraded by race mixing. In a line reflecting Chamberlain's assertion that the Jews' very existence was a "sin," Hitler declared that such racial mixing would be "a sin against the Will of the eternal Creator."[99] Not, it must be noted, a sin against the theory of Charles Darwin. "Regeneration" of the primitive German people and an elimination of blood poisoning can occur "so long as a fundamental stock of racially pure elements still exists and bastardization ceases."[100] Hitler thus sought a return to an ideal past, not an evolutionary advance to a transformed future.

## STRUGGLE FOR EXISTENCE

Most authors who try to connect Darwin with Hitler focus on Hitler's idea of "struggle," as if this implied Darwin's principle of "struggle for existence," that is, natural selection. The very title of Hitler's book, *My Battle* (or *Struggle, War*) hardly resonates of Darwinian usage—especially when one considers the title he originally planned: *A Four and a Half Year Battle [Kampf] against Lies, Stupidity and Cowardice*. A simple word count indicates that Hitler had a mania for the notion of struggle that no simple acquaintance with the idea in a scientific work could possibly explain. The term appears in one form or another some 266 times in the first 300 pages of the 800-page book: from the simple *Kampf* (struggle) to *Bekämpfung* (a struggle), *ankämpfen* (to fight), *Kampffeld* (field of struggle), *Kampfeslust* (joy of struggle), and so forth.

Darwin's principle of natural selection was, of course, used to explain the transmutation of species. But if someone like Hitler denies the transmutation and descent of species, then no matter what language he employs, the concept behind the language cannot be that of natural selection. Let me set aside for the moment this crucial objection to Hitler's supposed employment of Darwin's

97. Hitler, *Mein Kampf*, 313.
98. Ibid., 316.
99. Ibid., 314.
100. Ibid., 443.

device and examine the role of "struggle" in *Mein Kampf* and in his never-published *Zweites Buch* (Second book).

The phrase used in the German translation of the *Origin of Species* for "struggle for existence" is "Kampf um's Dasein."[101] Hitler uses that phrase, or one close to it, twice in *Mein Kampf.* Those two instances occur in an almost 800-page book in which some form of the word *Kampf* appears on almost every page; by sheer accident such a phrase might spill from the pen of an obsessed individual who seems to know hardly any other word. Yet those two instances do have a Darwinian ring. Both come in a context in which Hitler is worried about the apparent reduction in births in Germany due to lack of land. He deployed the term in an effort to justify annexing "unused" land to the east (e.g., Poland, Ukraine). His convoluted argument runs like this: if Germans stay within their own borders, then restraint on propagation will be necessary, and compassion will require that even the weak will be preserved; moreover, barbarians lacking culture but strong in determination will take the unused land; hence Germans, the bearers of culture, ought to appropriate the area needed for living (*Lebensraum*). Hitler's argument makes little sense from a Darwinian perspective. If living conditions became restricted within closed borders, it would be the more fit who would survive, whereas if conditions became relaxed by moving into an unoccupied and fruitful land, then the fit and the less fit (by some measure) ought to have fairly equal chances. Hence, from a Darwinian point of view, the conclusion ought to be just opposite to the one Hitler drew. Be that as it may, Hitler did argue that maintaining current borders allowed the weaker to survive "in place of the natural struggle for existence, which lets live only the strongest and healthiest."[102] He further observed that the Jews may have convinced the cultured Germans that mankind could play a trick on nature by developing land within Germany's borders, so that this will "make the hard struggle for existence [unerbittlichen Kampf ums Dasein] superfluous." His fundamental view is that "mankind becomes great through

101. Heinrich Georg Bronn was the first translator into German of Darwin's *Origin of Species*: *Über die Entstehung der Arten im Thier- und Pflanzen-Reich durch natürliche Züchtung, oder Erhaltung der vervollkommneten Rassen im Kampfe um's Daseyn,* trans. H. Bronn (Stuttgart: E. Schweizerbart'sche Verlagshandlung und Druckerei, 1860). The translation was slightly revised by Julius Victor Carus, who translated the fourth and subsequent editions of the *Origin* into German: *Über die Entstehung der Arten durch natürlichen Zuchtwahl oder die Erhaltung der begünstigten Rassen im Kampfe um's Dasein*, trans. J. Victor Carus (Stuttgart: E. Schweizerbart'sche Verlagshandlung und Druckerei, 1867). The Carus editions were standard in the early twentieth century.

102. Hitler, *Mein Kampf*, 145: "tritt an Stelle des natürlichen Kampfes um das Dasein, der nur den Allerstärksten und Gesündesten am Leben läßt."

eternal struggle—in eternal peace men come to nothing."[103] Most of these usages, with one interesting exception, as I'll specify in a minute, come almost verbatim from Chamberlain, not Darwin.

Struggle, battle [*Kampf*] formed the leitmotif of Hitler's considerations of human development, especially his own: from his strife-ridden efforts at forming a political movement to the anticipated battle to restore the German nation to world-historical standing. Like Wotan, he struggled against malicious dwarfs and thundering giants to obtain the ring of power, and for a brief historical moment, he succeeded. He even projected this struggle onto nature herself. In his never-published *Second Book*, he set out a brief prologue to his formulation of the National Socialist Party's foreign policy, a policy that outlined a political contest to restore German territory lost during the war, to expand the boundaries of the nation eastward, and even to recruit Italy and England as allies. In the prologue's brief creation myth, Hitler depicted the very forces of nature as struggling with each other to bring forth the earth: "The battle [Kampf] of natural forces with each other, the construction of a habitable surface of this planet, the separation of water and land, the formation of the mountains, the plains, and the seas."[104] One can almost hear the Wagnerian thunderbolts crashing. But immediately another distinctively German motif comes into play: human development became possible only after man began reflecting on his own history:

> World history [Weltgeschichte] in the period before the appearance of human beings was a representation of geological events. . . . Later, with the appearance of organic life, the interests of human beings became focused on the development and destruction of the many thousands of forms. And rather late man finally became visible to himself, and thus under the concept of world history [Weltgeschichte], he came to understand principally the history of his own becoming [seines eigenen Werdens], that is the representation of his own development [seiner eigenen Entwicklung zu verstehen]. This development is marked by an eternal struggle of men against animals and against other men. From the invisible chaos of individuals, endless structures, tribes, groups, peoples,

103. Ibid., 149.

104. Adolf Hitler, *Hitlers Zweites Buch* (Stuttgart: Deutsche Verlags-Anstalt, 1961), 46. Hitler dictated this statement of foreign policy in summer 1928; the publisher recommended against publishing since it would compete with the second volume of *Mein Kampf*, which at the time was not selling well. Later, in 1958, the manuscript was recovered from a U.S. Army deposit of confiscated papers.

FIGURE 9.6 Hitler at Bayreuth in 1938, with Winifred Wagner (1897–1980) and her son Wieland Wagner (1917–1966), daughter-in-law and grandson of Richard Wagner. (Courtesy of Getty Images)

and states finally arise, while the representation of their rise and fall is the depiction of an eternal struggle for life [eines ewigen Lebenskampfes]. If politics is history as it unfolds . . . then politics is in truth the continuation of the life struggle [Lebenskampfes] of a people.[105]

In this introductory passage to his *Second Book*, Hitler composed a libretto of second-hand Hegelian historicism accompanied by Wagnerian cries of incessant battle, of the unfolding of world history led by a Teutonic knight. Undoubtedly, as Alan Bullock has suggested, Hitler identified with one of Hegel's "world-historical individuals"—an Alexander, Caesar, or Napoleon—through whom the "will of the World-Spirit [Weltgeist]" was enacted.[106] In Hegel's view, man became gradually visible to himself only after he reflected on his historical character and slowly came to appreciate the evolution of world his-

105. Ibid., 47.
106. Alan Bullock, *Hitler: A Study in Tyranny*, abridged ed. (New York: Harper Perennial, 1991), 215.

tory (*Weltgeschichte*) according, as he put it, to "the principle of development [das Prinzip der Entwicklung]."[107] For Hegel as well as for Hitler, historical development entailed the unfolding of an ultimately rational process in which, according to Hegel, the "spirit is in a hard, ceaseless struggle [unendlicher Kampf] with itself."[108] Through a world historical figure like a Napoleon—or a Hitler—an inexorable destiny "develops," or evolves. Hegel, I presume it will be conceded, was no Darwinian.

Although Hegel emphasized the struggle that characterized world-historical events, Hitler's vision trembled with the fury of gods in constant battle, a vision that bears only superficial resemblance to Darwin's conception of species struggle. Before facile claims about a supposed identity are made, one needs examine the deeper sources of Hitler's argument and its goal. His general conception that humanity develops culturally through struggle and that racial mixing causes degeneration—these ideas replicate those of Chamberlain, who likewise signaled to his reader that "the idea of struggle governs my presentation [in *Foundations*]."[109] Chamberlain accepted Gobineau's contention that miscegenation caused cultural decline, but he insisted that such decline was not inevitable; one could struggle against degeneration and keep the Aryan folk, the bearers of culture, pure. The fight, however, had to be constantly renewed. "The struggle in which the weaker human material is eradicated [zu Grunde geht]," Chamberlain argued, "steels the stronger; moreover the struggle for life [Kampf ums Leben] strengthens the stronger by eliminating the weaker elements."[110] Hitler clearly echoed Chamberlain's observation that a peaceful land sows only cultural mediocrity; such a land, according to Chamberlain, "knows nothing of the social questions, of the hard struggle for existence [vom bittern Kampf ums Dasein]."[111] Compare this phrase with Hitler's "the hard struggle for existence [unerbittlichen Kampf ums Dasein]."[112] Hitler is thus not recycling Darwin but rather aping Chamberlain.[113] Neither

107. Georg Wilhelm Friedrich Hegel, *Vorlesungen über die Philosophie der Geschichte*, in *Werke*, vol. 12, ed. Eva Moldenhauer and Karl Michel, 4th ed. (Frankfurt am Main: Suhrkamp, 1995), 75: "The principle of *development* [Das Prinzip der *Entwicklung*] contains this as well, that an inner purpose [Bestimmung], a fundamental, intrinsic condition, establishes its own existence. This formal purpose is essentially the spirit that has world-history as its theater, its possession, and the field of its realization." It's hard to know whether Hitler read the *Vorlesungen* (Lectures) directly or derived the gist of Hegel's conception of history from some other source. That Hegel was Hitler's ultimate source, though, is unmistakable.

108. Ibid., 76.

109. Chamberlain, *Grundlagen*, 2:536.

110. Ibid., 1:277–78.

111. Ibid., 1:44.

112. Hitler, *Mein Kampf*, 149; also quoted previously.

113. Chamberlain, *Grundlagen*, 2:805.

Chamberlain nor Hitler conceived the goal of struggle to be the biological transformation of the German race into something different. Rather they thought means had to be taken to preserve the pure blood of the race and to realize, through struggle, the potential of the Teutons, who "alone have the ability for higher culture." The explicit purpose of the volkish state, according to Hitler was "the preservation of the racial element that supplies culture."[114] Thus, not transformation but preservation of the ancient race of the Germans.

It might be thought that I am quibbling about technicalities. Hitler after all used a phrase of Darwinian provenance, which points to the ultimate source of his ideas. But we are talking about ideas, not mere words, and the ideas that Hitler deploys are not Darwin's. If words alone are to be the criterion, one might just as easily ascribe his enthusiasm for struggle to Christianity, the greatness of which he explicitly identified with its constant struggle against other religions and its efforts to extirpate them.[115]

## THE POLITICAL SOURCE OF HITLER'S ANTI-SEMITISM

An obviously crucial question, concerning the supposed influence of Darwin on Hitler, is whether Darwinian concepts actually caused Hitler to adopt his racial ideas, especially his virulent anti-Semitism. I've already suggested the impact of Gobineau and Chamberlain (with a tincture of Hegel), but Hitler came to these more theoretical works with his anti-Jewish sentiments already in flower. Whence the beginnings, then, of his anti-Semitism?

In *Mein Kampf*, Hitler is perfectly explicit about the sources of his anti-Jewish attitudes. He identifies two political figures who turned him from an individual hardly aware of Jews into a passionate anti-Semite: Karl Lueger (1844–1910), newspaper baron and the mayor of Vienna (1897–1910); and Georg Schönerer (1842–1921), member of the Austrian Parliament and leader of the Pan-German Party, which sought to unite the German-speaking lands in a political confederation. Both were large presences in Vienna when Hitler, as an eighteen-year-old art student, arrived there from Linz in 1908. He claimed that before coming to the city he had little experience of Jews, thinking them merely Germans.[116] Vienna was awash in anti-Semitic pamphlets and broadsides, which he said were so exaggerated that he could hardly believe them. But Lueger and Schönerer made clear what was at stake in the Jewish question.

114. Hitler, *Mein Kampf*, 431, 434.
115. Ibid., 385, 506.
116. Ibid., 55.

FIGURE 9.7 Karl Lueger (1844–1910), mayor of Vienna (1897–1910), with Emperor Franz Joseph (1830–1916). Photo around 1905. (Courtesy of Österreichische Nationalbibliothek)

The Catholic Lueger was quite anti-Semitic mostly, it seems, for political advantage. When challenged on one occasion that his dinner companions were Jewish, he famously proclaimed: "I decide who's a Yid."[117] Lueger was opportunistic perhaps, but his newspaper, the *Volksblatt*, was so vehemently anti-Semitic that the archbishop of Vienna denounced it. Leuger's party shared both name and outlook with those of the Protestant court preacher and deeply anti-Semitic Adolf Stöcker. Hitler explicitly said that it was Lueger and his Christian Social Party that caused his "opinions regarding anti-Semitism to undergo a slow change in the course of time." "It was," he said, "my most serious change of opinion."[118]

Schönerer was even more anti-Semitic than Lueger, apparently from deep conviction rather than political opportunism. In *Mein Kampf*, Hitler compared Schönerer to Lueger: "At the time, Schönerer seemed to me the better and

117. Evans, *Coming of the Third Reich*, 43.
118. Hitler, *Mein Kampf*, 59.

FIGURE 9.8 Georg Schönerer (1842–1921), member of the Austrian Parliament and leader of the Pan-German Party. Photo about 1900. (Courtesy of Österreichische Nationalbibliothek)

more fundamental thinker in regard to the principal problems." As leader of the Pan-German Party, Schönerer sought a union of all German-speaking territories, a goal Hitler embraced as a young man. But, as Hitler recalled, he finally determined that Lueger was the sounder theorist of the two.[119] Hitler scholars

119. Ibid., 107.

Richard Evans and Ian Kershaw concur with Hitler's own estimate that these two politicians were the most significant in forming his attitudes about Jews and the need for a racially homogeneous German land.[120] So by Hitler's own admission, these political figures, not Darwin, were pivotal in forming his anti-Semitic attitudes. Thus neither was Hitler's conception of race Darwinian nor was Darwinism the source of his anti-Semitism. The motivation and origin of his views were political, not scientific, and certainly not Darwinian.[121]

## ETHICS AND SOCIAL DARWINISM

Although Hitler's conception of race was non-Darwinian, yet perhaps, somehow, his ethical views derived from Darwin, as Weikart's *Hitler's Ethic* urges. What was Darwin's ethical theory? That's not hard to determine, since he set it out explicitly in the *Descent of Man*. Darwin argued that human ethical behavior was rooted in social instincts of parental care, cooperation, and acting for the community welfare. These, as he formulated them, were altruistic instincts. Once protohumans had developed sufficient intelligence and memory to appreciate unrequited social instincts and once they began to speak and thereby could codify rules of behavior, then a distinctively human conscience would have emerged in the group. Early protohuman clans that had more altruists—that is, members who cooperated in providing for the general welfare and in food gathering and defense—would have an advantage over those with no or few altruists and would come to supplant them. Darwin further envisioned that while the concern of early humans would be their immediate communities, through the development of culture and science humans would come to view all men as their brothers, recognizing that the distinctions of skin color, head shape, and other racial traits were only superficial markers of a common humanity.[122] Darwin's conception of the widening circle of moral concern has nothing in common with Hitler's virulent hostility to races other than the Aryan. Moreover, since Darwin's theory is based on the emergence of

120. Evans, *Coming of the Third Reich*, 164–65; Ian Kershaw, *Hitler*, 2 vols. (New York: Norton, 2000), 1: 1–36. It may be that Hitler did have some knowledge of Jews while in Linz, but his attitude seemed to concretize, bathed as it was in the acidic opinions of Lueger and Schönerer.

121. Boyer is quite clear that Lueger's anti-Semitism had nothing to do with race but with political advantage. See John W. Boyer, *Karl Lueger (1844–1910), Christlichsoziale Politik als Beruf* (Vienna: Böhlau Verlag, 2010), 208.

122. These ideas are worked out in the Darwin, *Descent of Man* (1871), vol. 1, chaps. 3 and 5. I have discussed Darwin's ethical theory and its sources in Robert J. Richards, *Darwin and the Emergence of Evolutionary Theories of Mind and Behavior* (Chicago: University of Chicago Press, 1987), 185–242.

human groups from lower animals, it would be completely antithetic to Hitler's assumption of the permanency of races.

Any number of scholars who have written on the political and intellectual state of Germany in the 1930s and 1940s have described Hitler as advocating social Darwinism.[123] The term is quite vague. Indeed, it is often remarked that while Herbert Spencer might be a social Darwinist, Darwin himself was not. I believe one can discriminate some six traits that scholars usually have in mind when referring to social Darwinism:

1. The human races form a hierarchy from lower to higher, with the criteria for ranking being intelligence, morality, and cultural values.
2. Laws of nature apply equally to animals and men.
3. There is a struggle among human groups.
4. Knowing the laws of nature, humans can control the struggle to the advantage of the superior races.
5. The superior race is morally permitted to police its own group by eliminating the physically or intellectually inferior and promoting those of sound hereditary features.
6. The superior race may restrict the behavior of the lower races, even exterminating them.

The last two items, of course, give the category of social Darwinism its decidedly negative bite. I have not included the idea of transmutation of species, certainly a necessary feature of anyone who is also to be called a Darwinian simply. These six traits usually characterize most eugenicists working in the first part of the twentieth century. They also seem to capture Hitler's racism. Were they embraced by Darwin?

Before answering that last question, we might reflect that, after a fashion, these traits could also be applied, for instance, to Aristotle, who did not have moral qualms about slavery and who assumed the natural superiority of some groups of people. Likewise many slaveholders in the American South would likely have signed on to these propositions. Darwin did adopt propositions 1–4 but rejected 5 and 6. When he was confronted with the idea that it would be of long-term benefit for a society to prevent the weak in mind and body from marrying and reproducing their type, he demurred: "We must bear without complaining the undoubtedly bad effects of the weak surviving and propagating their kind." The attempt to check our sympathies for the poor and wretched

123. I have mentioned those recent scholars who casually employ the term "social Darwinism" in n. 8.

of the earth would, Darwin averred, cause "deterioration in the noblest part of our natures."[124] Of course, Hitler certainly followed all of the precepts, including 5 and 6. So while convention might sanction calling Hitler a social Darwinian (even if he did not believe in species transmutation), that same convention could not be applied to Darwin himself. Thus the name "social Darwinian" is misleading and itself should imply no connection with the ethical theory of Charles Darwin.

Hitler rejected the transmutation of species and instead held to the older notion of fixity of type; he deployed notions of struggle between races but derived the idea from non-Darwinian sources; and if he were to be called a social Darwinian, that same designation with its intended meaning could not also describe Darwin's views. Hitler's anti-Semitism, as he himself avowed, stemmed from political not scientific sources. Consequently no reasonable evidence links Hitler's racial dogmas to Darwin's theory. Despite this conclusion, one might still contend that while Hitler did not personally derive ideas from Darwin, he fostered a scientific regime that made Darwinism and Haeckelianism the chief arbiters in questions of race.

## WAS THE BIOLOGICAL COMMUNITY DARWINIAN UNDER HITLER?

The answer to the question of whether the biological community during the Nazi period was Darwinian is complicated by this salient fact: many extremely good scientists remained in Germany during the Nazi period and practiced science at a very high level. One has only to mention the names of Werner Heisenberg (1901–1976) and Werner von Braun (1912–1977) to recognize that, despite their politics, they were extraordinary scientists. In biology likewise, some exceedingly good biologists of different theoretical orientations could be found in the universities and research institutes of Nazi Germany. For instance, the Nobel Prize winner (1969) Max Delbrück (1906–1981) worked in biophysics in Berlin during the early part of Hitler's regime before receiving a research fellowship for work in the United States in 1937 and never returning to Nazi Germany; his great colleague Nikolai Vladimirovich Timoféeff-Ressovsky (1900–1981) continued as director of the genetics division of the Kaiser Wilhelm Institute for Brain Research through the end of the war.[125]

124. Darwin, *Descent of Man* (1871), 1:169, 168–69.

125. For an account of Timoféeff's career, see Vadim Ratner, "Nikolay Vladimirovich Timoféeff-Ressovsky (1900–1981): Twin of the Century of Genetics," *Genetics* 158 (2001): 933–39; and Yakov Rokityanskij, "N V Timofeeff-Ressovsky in Germany (July 1925–September 1945)," *Journal of Biosciences* 30,

Many topflight biologists, some of whom were Darwinians, remained in Germany while Hitler was in power. Of course, many others connected with the regime were non-Darwinians and, by any standards, quite awful. During the late 1930s and 1940s, the discipline of biology itself underwent a significant transition. Initially, through the 1910s and 1920s, Mendelian genetics and Darwinian natural selection theory were often regarded as opposed, with the former considered to be real science and the latter romantic butterfly collecting.[126] During the next two decades, however, biologists discovered their complementary features; as a result, Mendelian genetics and Darwinian evolutionary theory became joined in the synthetic framework that now serves as the foundation of modern biological science. Several German biologists of the period contributed to this development, though others retained the older attitude. Without doubt, then, Darwinian evolutionary biologists worked in Germany during the Hitler period. And some Darwinians, such as the Tübingen botanist Ernst Lehmann (1880–1957), founder (1931) of the Association of German Biologists and its journal *Der Biologe*, argued for a distinctively German biology aligned with the goals of the Nazi party.[127] The pertinent question, though, is whether the National Socialist Party gave special accord to Darwinian science. In 1940, the year he took up a professorship at Königsberg, Konrad Lorenz (1903–1989), good Darwinian that he was, complained that there were many "in the schools of National-Socialistic greater Germany who in fact still reject evolutionary thought and descent theory [Entwicklungsgedanken und Abstammungslehre] as such."[128] Lorenz's complaint strongly implies that Darwinism had no official mandate in the educational system. Even more compelling evidence can be drawn from an examination of a leading scientific journal of the period that was also an official organ of the Nazi party, *Zeitschrift für die gesamte Natur-*

no. 5 (2005): 573–80. See also Kristie MacRakis, *Surviving the Swastika: Scientific Research in Nazi Germany* (Oxford: Oxford University Press, 1993), 120–22.

126. The geneticist and formidable historian of biology of the first part of the twentieth century, Erik Nordenskiöld, declared in 1928 that Darwin's theory "has long ago been rejected in its most vital points by subsequent research." It would be replaced, he thought, by real science, modern laboratory genetics. See Erik Nordenskiöld, *The History of Biology*, trans. L. B. Eyre (New York: Tudor, [1928] 1936), 477.

127. See Ernst Lehmann, *Biologie im Leben der Gegenwart* (Munich: J. F. Lehmann Verlag, 1933), 212–38. Lehmann attempted to show that modern evolutionary biology, with the important addition of Mendelism, aligned perfectly with goals of Hitler and his party. His main concern, in so far as biology was to serve the state, was to warn of the dangers of racial decline through hybridization with lower races (216–23). Though Lehmann tried several times to join the Nazi Party, he was always rejected, ultimately because he fell afoul of more powerful party leaders. See Ute Deichmann's discussion of Lehmann's plight in Deichmann, *Biologists under Hitler*, trans. Thomas Dunlap (Cambridge: Harvard University Press, 1996), 74–89.

128. Konrad Lorenz, "Nochmals: Systematik und Entwicklungsgedanken im Unterricht," *Der Biologe* 9 (1940): 24–36, quotation at 24.

*wissenschaft* (Journal for all of natural science), which published from 1935 to 1944. From its third year, the journal carried the subtitle: "Organ of the Natural Science's Professional Division of the Reich's Student Leadership."

The *Zeitschrift* published articles principally in the physical sciences and biology, along with essays on philosophical treatments of those sciences. It sought to purge scientific activity of Jewish influences and establish Aryan science free from alien taint.[129] On one marked occasion in the journal's pages, Werner Heisenberg had to defend modern physics—particularly relativity theory and quantum theory—from charges that it was incompatible with National Socialism.[130] The journal published in all areas of biology, but with particular concern for the field's relationship to the ideology of National Socialism.

The tone and attitude of the journal were established in the first article of the first volume (1935) by a philosopher from Kiel, Kurt Hildebrandt (1881–1966), who was also an editor.[131] In "Positivismus und Natur," Hildebrandt responded to an article published by the quantum physicist Pascual Jordan (1902–1980), who claimed that positivism was the method of all science. Jordan argued that both the subjective world of consciousness and the objective world of nature could be derived from neutral experience without any appeal to metaphysics.[132] Hildebrandt objected that this really reduced consciousness to mechanism and failed to recognize that natural phenomena depended on a creative spirit, of the sort suggested by both Goethe and Nietzsche. "What is

129. When the journal became an official party organ in 1937, a new editorial indicated that the journal took as its task "the cultivation of scientific content in so far as it reflects an essential German nature." See editorial, *Zeitschrift für die gesamte Naturwissenschaft* 3 (1937–38): 1. Deichmann discusses the character of the journal in Deichmann, *Biologists under Hitler*, 43.

130. Werner Heisenberg, "Die Bewertung der modernen theoretischen Physik, *Zeitschrift für die gesamte Naturwissenschaft* 9 (1943): 201–12. Heisenberg rejected the idea of the incompatibility of modern physics and National Socialism (210–11). He noted that his essay had been written in 1940, which was about the time a fight occurred over who should fill the chair held by the retiring physicist at Munich Arnold Sommerfeld. Heisenberg and other students of Sommerfeld tried to prevent the group supporting *Deutsche Physik*—which was anti-Semitic and hostile to relativity and quantum mechanics—from advancing their candidate to the chair. Heisenberg was the heir apparent—having won the Nobel Prize in 1932—yet he lost the fight. Nonetheless his stature grew as the possibility of a nuclear weapon was considered. In 1943, when his paper was published, he had been appointed to the chair of theoretical physics at the University of Berlin and made a member of the Prussian Academy of Sciences.

131. Kurt Hildebrandt, "Positivismus und Natur," *Zeitschrift für die gesamte Naturwissenschaft* 1 (1935–36): 1–22. Martin Heidegger was one of the associate editors (*Mitarbeiter*) of the journal.

132. See Pascual Jordan, "Über den positivistischen Begriff der Wirklichkeit," *Die Naturwissenschaften* 22 (20 July 1934): 33–39. Jordan contended that experience alone was the foundation for science and that it united the subjective world and the objective world. Not only did Hildebrandt reject the analysis, but so did many members of the Vienna Circle, particularly Otto Neurath. See the discussion of this dispute within the movement of logical positivism in Suzanne Gieser, *The Innermost Kernel: Depth Psychology and Quantum Physics, Wolfgang Pauli's Dialogue with C. G. Jung* (New York: Springer, 2005), esp. 50–102.

called positivism today, worse than any older philosophy that went under that name, denies actual spiritual experience." This is shown, he thought, especially in the opposition of French rationalism and English empiricism to the notion of "creative spirit" (*schöpferische Geist*): "German nature-philosophy found in Leibniz, Herder, and Goethe showed the correct way to overcome this opposition by proposing a union of spirit and matter, which as a world view is most graphically expressed by the term 'pantheism.' In respect of creative nature as development, Leibniz already had a theory of species descent [Abstammungstheorie]."[133]

Hildebrandt thus thought that English biology of the nineteenth century was inadequately grounded, but now "exact biology has dealt Darwin's mechanization a deathblow [Todesstoß]." He claimed that the new theory of inheritance, "which had long been suppressed by Darwinism, has had unexpected success." Darwinism, according to Hildebrandt, had to be rejected: "the creative unfolding of species, the origin of species from the amoeba to man, cannot be explained by this mechanistic theory. Rather exact research on heritability has clearly destroyed the mechanistic framework of Darwinian theory." What exactly Hildebrandt meant by "creative spirit," "creative force," and the like—or the new research in genetics—is not at all clear in his essay. In a footnote to the passage I've just quoted he added: "This is not a reference to vitalism. Goethe and Schelling were not vitalists, but monists, since they recognized the same creative power in the universe as in living individuals; they were the opponents of empiricism and materialism, which agree with conventional belief in God."[134] By the new theory of inheritance he likely meant that associated with Hugo de Vries's (1848–1935) mutation theory, which supposed that macromutations, not Darwinian gradualism, led to the appearance of new species.[135] But Hildebrandt also suggested, despite disavowals, that there was a definite sort of élan vital behind such transitions. Volume 4 of the *Zeitschrift* carried a long article by Hans Driesch (1867–1941), who also supposed that species change could not be explained by any Darwinian or Haeckelian mechanistic process but required the postulation of a vital entelechy of the sort conceived by Aristotle.[136] All of this, of course, is antithetic to Darwinism.

133. Hildebrandt, "Positivismus und Natur," 20, 21.

134. Ibid., 22.

135. That he had de Vries's theory in mind seems fairly clear from a subsequent article of his in the journal: Kurt Hildebrandt, "Die Bedeutung der Abstammungslehre für die Weltanschauung," *Zeitschrift für die gesamte Naturwissenschaft* 3 (1937–38): 15–34.

136. Hans Driesch, "Der Weg der Theoretischen Biologie," *Zeitschrift für gesamte Naturwissenschaft* 4 (1938–39): 209–32.

When the *Zeitschrift* became an official organ of the National Socialist Party, it did not change its orientation, nor did Hildebrandt. In volume 3 (1937–38), he proclaimed: "Our modern theory of inheritance has not supported this hypothesis [of descent], but endangers the foundational assumptions of Darwin and Haeckel. Mendelian research rests on the assumption of an unchanging species and mutation-theory has, indeed, several disadvantages, but does not attempt to explain or demonstrate the origin of a higher species." He darkly hinted that "real transmutation theory cuts across, if ever so carefully, the border to metaphysics."[137]

One of the new editors of the *Zeitschrift* after the political *Gleichschaltung* (takeover) by the Nazi party, the botanist Ernst Bergdolt (1902–1948), contended that the Darwinian selection principle was typical of the kind of passive environmentalist theory declaimed by Jewish liberals.[138] In a dispute between a Darwinian and an anti-Darwinian anthropologist, Bergdolt lent his editorial support to the latter.[139] The Darwinian, Christian von Krogh (1909–1992) of Munich, argued that Haeckel's scheme of human descent from apelike forbearers had evidence on its side,[140] while the anti-Darwinian, Max Westenhöfer (1871–1957) of Berlin, drew from comparative anatomy the opposite conclusion. Westenhöfer, as a student of Rudolf Virchow, declared that "from numerous comparative-morphological investigations during the last twenty years, I came, almost against my will, to a critical rejection of the Darwin-Haeckel doctrine and was forced to construct a new theory of the heritage of mankind."[141] Westenhöfer adopted a version of de Vries's mutation theory to explain human development through a lineage independent of the ape-man hypothesis.

Writing in the *Zeitschrift* after it became a party organ in 1937, Günther Hecht (1902–1945), an official of the party's Department of Race Policy (Rassenpolitischen Amt der NSDAP) and member of the Zoological Institute in Berlin, completely rejected the idea (*grundsätzlich abgelehnt*) that the materialistic theories of Darwin and especially Haeckel had anything to do with

137. Hildebrandt, "Die Bedeutung der Abstammungslehre, 22.

138. Ernst Bergdolt, "Zur Frage der Rassenentstehung beim Menschen," *Zeitschrift für die gesamte Naturwissenschaft* 3 (1937–38): 109–13.

139. Ernst Bergdolt, "Abschließende Bermerkungen zu dem Thema 'Das Problem der Menschenwerdung," *Zeitschrift für die gesamte Naturwissenschaft* 6 (1940): 185–88.

140. Christian von Krogh, "Das 'Problem' Menschenwerdung," *Zeitschrift für die gesamte Naturwissenschaft* 6 (1940): 105–12. Uwe Hoßfeld provides a brief account of Krogh's position in Hoßfeld, *Geshichte der biologischen Anthropologie in Deutschland* (Stuttgart: Franz Steiner Verlag, 2005), 272–74.

141. Max Westenhöfer, "Kritische Bemerkung zu neueren Arbeiten über die Menschenwerdung und Artbildung," *Zeitschrift für die gesamte Naturwissenschaft* 6 (1940): 41–62, quotation at 41.

the "völkisch-biological position of National Socialism."[142] The head of the Department of Race Policy, the physician Walter Groß (1904–1945), thought the party ought to remain clear of any commitment to the doctrines of human evolution, "which is frequently still pervaded with Haeckelian ways of thinking in its basic ideological ideas . . . and is thus publicly considered a part of materialistic, monist ideas."[143]

The rejection of Haeckelian ideas had been sealed in 1935 when the Saxon ministries of libraries and bookstores banned all material inappropriate for "National-Socialist formation and education in the Third Reich." Among the works to be expunged were those by "traitors," such as Albert Einstein; those by "liberal democrats," such as Heinrich Mann; literature by "all Jewish authors no matter what their sphere"; and materials by individuals advocating "the superficial scientific enlightenment of a primitive Darwinism and monism," such as Ernst Haeckel.[144] It is quite clear that Darwinian evolutionary theory held no special place within the community of biologists supportive of National Socialism. Rather, biologists and philosophers most closely identified with the goals of the Nazi party and officials in that party utterly rejected Darwinian theory, especially as advanced by Darwin's disciple Ernst Haeckel.

Weikart and others have found the poison within the tempting apple of Darwinian theory to be its materialism, the feature that, according to Weikart, led to the pernicious morality of Hitler and his Nazi biologists. But leading Nazi biological theorists not only rejected Darwinism but they did so precisely because of its supposed materialism. Could there be anything left of the claim that Hitler derived his racial attitudes from Darwinian theory?

## CONCLUSION

Countless conservative religious and political tracts have attempted to undermine Darwinian evolutionary theory by arguing that it was endorsed by Hitler and led to the biological ideas responsible for the crimes of the Nazis. These dogmatically driven accounts have been abetted by more reputable scholars who have written books with titles such as *From Darwin to Hitler*. Ernst Haeckel, Darwin's great German disciple, is presumed to have virtually packed

142. Günther Hecht, "Biologie und Nationalsozialismus," *Zeitschrift für die gesamte Naturwissenschaft* 3 (1937–38): 280–90, quotation at 285.

143. Walter Groß, as quoted in Deichmann, *Biologists under Hitler*, 270.

144. "Richtilinien für die Bestandsprüfung in den Volksbüchereien Sachsens," Die Bücherei 2 (1935): 279–80.

his sidecar with Darwinian theory and monistic philosophy and delivered their toxic message directly to Berchtesgaden—or at least, individuals such as Daniel Gasman, Stephen Jay Gould, and Larry Arnhardt have so argued. In this chapter I have maintained that these assumptions simply cannot be sustained after a careful examination of the evidence.

To be considered a Darwinian, one must endorse at least three propositions: that the human races exhibit a hierarchy of more advanced and less advanced peoples; that over long periods of time, species have descended from other species, including the human species, which derived from apelike ancestors; and that natural selection—as Darwin understood it—is the principle means by which transmutation occurs. Hitler and the Nazi biologists I have considered certainly claimed a hierarchy of races, but that idea far antedated the publication of Darwin's theory and was hardly unique to it. There is no evidence linking Hitler's presumption of such a hierarchy and Darwin's conception. Moreover, Hitler explicitly denied the descent of species, utterly rejecting the idea that Aryan man descended from apelike predecessors. And most of the Nazi scientists I have cited likewise rejected that aspect of Darwin's theory. Hitler did speak of the "struggle for existence" but likely derived that language from his friend and supporter Houston Stewart Chamberlain, an avowed anti-Darwinian. By Hitler's own testimony, his anti-Semitism had political, not scientific or biological roots; there is no evidence that Hitler had any special feeling for these scientific questions or read anything Darwin wrote. Among Nazi biologists, at least those publishing in an official organ of the party, Mendelian genetics and de Vriesian mutation theory were favored, with both vying at the beginning of the twentieth century to replace Darwinian theory. The perceived mechanistic character of Darwinism stood in opposition to the vitalistic conceptions of Nazi biologists and that of Hitler—or at least vitalism resonated more strongly with Hitler's thoughts about race. Moreover, although his own religious views remain uncertain, Hitler often enough assumed a vague theism of a sort usually pitted against Darwinian theory.

If the term "social Darwinian" refers to individuals who apply evolutionary theory to human beings in social settings, there is little difficulty in denominating Herbert Spencer or Ernst Haeckel a social Darwinian. With that understanding, Darwin himself also would have to be so called. But how could one possibly ascribe that term at the same time to Hitler, who rejected evolutionary theory? Only in the very loosest sense, when the phrase has no relationship to the transmutational theory of Charles Darwin or Darwin's particular ethical views, might it be used for Hitler.

As I suggested at the beginning of this chapter, there is an obvious sense in which my claims must be moot. Even if Hitler could recite the *Origin of Species* by heart and referred to Darwin as his scientific hero, that would not have the slightest bearing on the validity of Darwinian theory or the moral standing of its author. The only reasonable answer to the question that gives this chapter its title is a very loud and unequivocal No.

# ACKNOWLEDGMENTS

President Barack Obama, once a member of my university, offered the salutary admonition to the business community that while many showed ingenious enterprise, they did not achieve their successes alone but received the help of others—relatives and friends, as well as governmental institutions that established structures necessary for their operations to succeed. Any scholar who has achieved even a modicum of success must likewise acknowledge that he or she could not have gone it alone but was nurtured and sustained by relatives, colleagues, and scholarly institutions. My colleagues and students at the University of Chicago offered a measured elixir of encouragement and challenge that made the essays in this volume possible. During the period of their composition, I also enjoyed the support of several other institutions: the Max Planck Institute for the History of Science, the National Science Foundation, and the Guggenheim Foundation. The following libraries and archives allowed me to quote from letters and other manuscript material: Department of Manuscripts, Cambridge University Library; Senate House Library, London; Haeckel-Haus, Jena; and Bristol University Library.

Several individuals have had a more direct hand in offering advice concerning the arguments made in the various essays of this volume, and I acknowledge their help and particularly their critical misgivings: Jerry Coyne, Christopher DiTeresi, Matthias Dörries, Michael Geyer, Christopher Haufe, Jonathan Hodge, David Kohn, Alessandro Pajewski, Trevor Pearce, and Gregory Radick. I owe a special debt to Michael Ruse, who provided a perfect foil for many of the positions I have taken in these essays. Lily Huang carefully read the manuscript and furnished invaluable advice on both style and argument.

My editor, Karen Merikangas Darling, as usual, guided me through the process and added the critical word at the right moment. My wife Barbara is she without whom nothing is possible for me.

I am grateful for permission to publish essays in revised form that previously appeared in the following publications: *Cambridge Companion to the Origin of Species*, Cambridge University Press (chap. 2); *Proceedings of the National Academy of Sciences* (chap. 2); *Studies in the History and Philosophy of Biology and Biomedical Sciences* (chap. 3); *Cambridge Companion to Charles Darwin*, Cambridge University Press (chap. 4); *Herbert Spencer: The Intellectual Legacy*, Galton Institute (chap. 5); *Darwin: Art and the Search for Origins*, Shirn Kunsthalle (chap. 6); *Biology and Philosophy* (chap. 7); and *Experimenting in Tongues: Studies in Science and Language*, Stanford University Press (chap. 8).

# BIBLIOGRAPHY


Alberts, B., D. Bray, J. Lewis, M. Raff, K. Roberts, and J. Watson. *Molecular Biology of the Cell*. 3rd ed. New York: Garland, 1994.

Alexander, Richard. *The Biology of Moral Systems*. New York: Aldine De Gruyter, 1987.

Allen, Grant. Letter to Herbert Spencer (10 November 1874). MS 791, no. 102. Athenaeum Collection of Spencer's Correspondence, University of London Library.

Alter, Stephen. *Darwinism and the Linguistic Image*. Baltimore: Johns Hopkins University Press, 1999.

Arendt, Hannah. *The Origins of Totalitarianism*. Orlando, FL: Harcourt, 1994. First published 1948.

Arnhart, Larry. *Darwinian Conservatism*. Charlottesville, VA: Imprint Academic, 2005.

Arsleff, Hans. *From Locke to Saussure: Essays on the Study of Language and Intellectual History*. Minneapolis: University of Minnesota Press, 1982.

Bahr, Hermann. "Ernst Haeckel." In *Der Antisemitismus: Ein internationals Interview*. Berlin: S. Fischer, 1894.

Bain, Alexander. Letter to Herbert Spencer (17 November 1863). MS 791, no. 67. Athenaeum Collection of Spencer's Correspondence, University of London Library.

Beddall, Barbara. "Darwin and Divergence: The Wallace Connection." *Journal of the History of Biology* 21 (1988): 1–68.

Bell, Charles. *Expression: Its Anatomy and Philosophy*. 3rd ed. New York: Wells, 1873. First published 1844.

Bergdolt, Ernst. "Abschließende Bermerkungen zu dem Thema 'Das Problem der Menschenwerdung.'" *Zeitschrift für die gesamte Naturwissenschaft* 6 (1940): 185–88.

———. "Zur Frage der Rassenentstehung beim Menschen." *Zeitschrift für die gesamte Naturwissenschaft* 3 (1937–38): 109–13.

Biddiss, Michael. *Father of Racist Ideology: The Social and Political Thought of Count Gobineau.* New York: Weybright and Talley, 1970.

Bloch, Marc. *The Historian's Craft.* Translated by Peter Putnam. New York: Vintage Books, 1953.

Block, Ned, and Philip Kitcher. "Misunderstanding Darwin: Natural Selection's Secular Critics Get It Wrong." *Boston Review*, March–April 2010, 29–32.

Blumenbach, Johann Friedrich. *De Generis humani varietate nativa liber.* 3rd ed. Göttingen: Vandenhoek et Ruprecht, 1795.

Bölsche, Wilhelm. *Ernst Haeckel: Ein Lebensbild.* Berlin: Georg Bondi, 1909.

Bopp, Franz. *Vergleichende Grammatik des Sanskrit, Zend, Griechischen, Lateinischen, Litthauischen, Gothischen und Deutschen.* Berlin: Königlichen Akademie der Wissenschaften, 1833.

Bowler, Peter. *Darwin Deleted: Imagining a World without Darwin.* Chicago: University of Chicago Press, 2013.

———. *The Non-Darwinian Revolution: Reinterpreting a Historical Myth.* Baltimore: Johns Hopkins University Press, 1988.

———. *Theories of Human Evolution.* Baltimore: Johns Hopkins University Press, 1986.

———. "What Darwin Disturbed: The Biology That Might Have Been." *Isis* 99 (2008): 560–67.

Boyer, John. *Karl Leuger (1844–1910), Christlichsoziale Politik als Beruf.* Vienna: Böhlau Verlag, 2010.

Bradley, F. H. Letter to C. Lloyd Morgan (16 February 1895). DM 612. Papers of C. Lloyd Morgan, Bristol University Library.

Bronn, Heinrich Georg. "On the Laws of Evolution of the Organic World during the Formation of the Crust of the Earth." *Annals and Magazine of Natural History*, 3rd ser., 4 (1859): 81–90, 175–84.

———. "Schlusswort des Übersetzers." In Charles Darwin, *Über die Entstehung der Arten im Thier- und Pflanzen-Reich durch natürliche Zütung, oder Erhaltung der vervollkommneten Rassen in Kampfe um's Dasyn*, translated by H. G. Bronn, based on the 2nd English ed., 495–520; 525–51 in 2nd German ed. Stuttgart: Schweizerbart'sche Verhandlung und Druckerei, 1860 and 1863.

Brooke, John. "'Laws Impressed on Matter by the Creator'? The Origin and the Question of Religion." In *The Cambridge Companion to the "Origin of Species,"* edited by Michael Ruse and Robert J. Richards, 256–74. Cambridge: Cambridge University Press, 2009.

Brougham, Henry Lord. *Dissertations on Subjects of Science concerned with Natural Theology: Being the Concluding Volumes of the New Edition of Paley's Work.* London: Knight, 1839.

Browne, Janet. *Charles Darwin: The Power of Place.* New York: Alfred Knopf, 2002.

———. *Charles Darwin: Voyaging.* Princeton: Princeton University Press, 1995.

———. "Darwin's Botanical Arithmetic and the 'Principle of Divergence,' 1852–1858." *Journal of the History of Biology* 13 (1980): 53–89.

Bullock, Alan. *Hitler: A Study in Tyranny.* Abridged ed. New York: Harper Perennial, 1991.

Butler, H., and B. Juurlink. *An Atlas for Staging Mammalian and Chick Embryos.* Boca Raton, FL: CRC Press, 1987.

Candolle, Alphonse de. *Géographie botanique raisonnée ou exposition des faits principaux et des lois concernant la distribution géographique des plantes de l'époque actuelle.* 2 vols. Paris: Librairie de Victor Masson, 1855.

Carus, Carl Gustav. *Denkschrift zum hundertjährigen Geburtsfeste Goethe's. Ueber ungleiche Befähigung der verschiedenen Menschheitstämme für höhere gestige Entwickelung.* Leipzig: Brockhaus, 1849.

———. "The Kingdoms of Nature, Their Life and Affinity." *Scientific Memoirs* 1 (1837): 223–54.

———. *System der Physiologie für Naturforscher und Aerzte.* 2 vols. Dresden: Gerhard Fleischer, 1838.

Chamberlain, Houston Stewart. *Cosima Wagner und Houston Stewart Chamberlain im Briefwechsel, 1888–1908.* Edited by Paul Pretzsch. Leipzig: Philipp Reclam, 1934.

———. *Das Drama Richard Wagner's. Eine Anregung.* Vienna: Breitkopf und Härtel, 1892.

———. *Die Grundlagen des neunzehnten Jahrhunderts.* 2 vols. Munich: Bruckmann, 1899.

———. *Houston Stewart Chamberlain, Auswahl aus seinen Werken.* Edited by Hardy Schmidt. Breslau: Ferdinand Hirt, 1935.

———. *Houston Stewart Chamberlain Briefe, 1882–1924, und Briefwechsel mit Kaiser Wilhelm II.* Edited by Paul Pretzsch. 2 vols. Munich: Bruckmann, 1928.

———. *Lebenswege meines Denkens*. Munich: Bruckmann, 1919.

———. *Natur und Leben*. Edited by J. von Uexküll. Munich: Bruckmann, 1928.

———. *Parsifal-Märchen*. Munich: Bruckmann, 1900.

———. *Recherches sur la sève ascendante*. Neuchâtel: Attinger Frères, 1897.

———. *Richard Wagner*. Munich: Bruckmann, 1896.

———. *Richard Wagner. Echte Briefe an Ferdinand Praeger*. Bayreuth: Grau'sche Buchhandlung, 1894.

Collins, P. "Embryology and Development." In *Gray's Anatomy*, 38th ed., edited by P. Collins, 91–341. London: Churchill Livingstone, 1995.

Cooke, Kathy J. "Darwin on Man in the *Origin of Species*." *Journal of the History of Biology* 26 (1990): 517–21.

Coyne, Jerry, and H. Allen Orr. *Speciation*. Sunderland, MA: Sinauer Associates, 2004.

Cuvier, Georges. *Le régne animal*. 2nd ed. 5 vols. Paris: Deterville Libraire, 1829–30.

Daston, Lorraine, and Peter Galison. *Objectivity*. New York: Zone Books, 2007.

Darwin, Charles. *The Autobiography of Charles Darwin*. Edited by Nora Barlow. New York: Norton, 1969.

———. *Charles Darwin's* Beagle *Diary*. Edited by R. D. Keynes. Cambridge: Cambridge University Press, 1988.

———. *Charles Darwin's Natural Selection: Being the Second Part of His Big Species Book Written from 1856 to 1858*. Edited by R. C. Stauffer. Cambridge: Cambridge University Press, 1975.

———. *Charles Darwin's Notebooks, 1836–1844*. Edited by Paul H. Barrett, Peter J. Gautry, Sandra Herbert, David Kohn, and Sidney Smith. Ithaca: Cornell University Press, 1987.

———. *The Correspondence of Charles Darwin*. Edited by Frederick Burkhardt et al. 19 vols. to date. Cambridge: Cambridge University Press, 1985–.

———. Letter to Caroline Kennard (9 January 1882). DAR 185. Department of Manuscripts, Cambridge University Library.

———. *The Descent of Man and Selection in Relation to Sex*. 2 vols. London: Murray, 1871.

———. *The Descent of Man and Selection in Relation to Sex*. 2nd ed. With an introduction by James Moore and Adrian Desmond. London: Penguin Group, 2004. First published 1879.

———. *The Expression of the Emotions in Man and Animals*. Chicago: University of Chicago Press, 1965. First published 1872.

———. *Foundations of the Origin of Species: Two Essays Written in 1842 and 1844*. Edited by Francis Darwin. Cambridge: Cambridge University Press, 1909.

———. Humble bee notebook. DAR 194.1–12. Department of Manuscripts, Cambridge University Library.

———. *Journal of Researches into the Geology and Natural History of the Various Countries Visited by H.M.S.* Beagle *under the Command of Captain FitzRoy, R.N., from 1832 to 1836*. London: Henry Colburn, 1839.

———. *Life and Letters of Charles Darwin*. Edited by Francis Darwin. 2 vols. New York: D. Appleton, 1891.

———. Loose notes. DAR 76.1–4. Department of Manuscripts, Cambridge University Library.

———. *Living Cirripedia, a Monograph on the Sub-class Cirripedia, with Figures of All the Species*. Vol. 1, *The Lepadidæ; or, Pedunculated Cirripedes*; vol. 2, *The Balanidæ, (or Sessile Cirripedes); the Verruciae*. London: Ray Society, 1852 and 1854.

———. Manuscript notes. DAR 205.3.167 and DAR 205.5.171. Department of Manuscripts, Cambridge University Library.

———. *A Monograph on the Fossil Balanidae and Verrucidae of Great Britain*. London: Printed for the Palaeontographical Society, 1854.

———. *A Monograph on the Fossil Lepadidae, or, Pedunculated Cirripedes of Great Britain*. London: Printed for the Palaeontographical Society, 1851.

———. *On the Origin of Species*. London: Murray, 1859.

———. *The Origin of Species: A Variorum Text*. Edited by Morris Peckham. Philadelphia: University of Philadelphia Press, 1959.

———. Personal journal. MS 34, DAR 158.1–76. Department of Manuscripts, Cambridge University Library.

———. *Über die Entstehung der Arten im Thier- und Pflanzen-Reich durch natürliche Züchtung, oder Erhaltung der vervollkommenten Rassen im Kampfe um's Daseyn*. Translated by J. Victor Carus. Stuttgart: E. Schweizerbart'sche Verlagshandlung und Druckerei, 1867.

———. *The Variation of Animals and Plants under Domestication*. 2 vols. London: Murray, 1868.

———. *The Variation of Animals and Plants under Domestication*. 2nd ed. 2 vols. New York: D. Appleton, 1899.

———. *The Voyage of the* Beagle. Edited by Leonard Engel. 2nd ed. New York: Doubleday, 1962. First published 1844.

Darwin, Charles, and Alfred Russel Wallace. "On the Tendency of Species to Form Varieties, and on the Perpetuation of Varieties and Species by Natural

Means of Selection." *Journal of the Proceedings of the Linnean Society of London*, Zoology 3 (1858): 45–62.

"The Darwinian Theory in Philology." *Reader* 3 (1864): 261–62.

Dawkins, Richard. *The Selfish Gene*. Oxford: Oxford University Press, 1976.

Deacon, Terrence. *The Symbolic Species: The Co-Evolution of Language and the Brain*. New York: Norton, 1997.

Deichmann, Ute. *Biologists under Hitler*. Translated by Thomas Dunlap. Cambridge: Harvard University Press, 1996.

Dennett, Daniel. *Darwin's Dangerous Idea: Evolution and the Meaning of Life*. New York: Simon and Schuster, 1995.

———. "Fun and Games in Fantasyland." *Mind and Language* 23 (2008): 25–31.

Desmond, Adrian, and James Moore. *Darwin's Sacred Cause*. New York: Houghton Mifflin Harcourt, 2009.

Dietze, Joachim. *August Schleicher als Slawist: Sein Leben und sein Werk in der Sicht der Indogermanistik*. Berlin: Academic-Verlag, 1966.

Dobzhansky, Theodosius. *Genetics and the Origin of Species*. With an introduction by Stephen Jay Gould. New York: Columbia University Press, 1982. First published 1937.

Driesch, Hans. "Der Weg der Theoretischen Biologie." *Zeitschrift für die gesamte Naturwissenschaft* 4 (1938–39): 209–32.

Dubois, Eugene. *Pithecanthropus erectus: Eine Menschenaenliche Uebergangsform aus Java*. Batavia: Landsdruckeri, 1894.

Duncan, David. *Life and Letters of Herbert Spencer*. 2 vols. New York: D. Appleton, 1908.

Eaton, John M. *A Treatise on the Art of Breeding and Managing Tame, Domesticated, Foreign and Fancy Pigeons*. London: printed by author, 1858.

Evans, Richard. *The Coming of the Third Reich*. New York: Penguin, 2003.

———. *The Third Reich in Power*. New York: Penguin, 2005.

Farrar, Frederick. "Philology and Darwinism." *Nature* 1 (1870): 527–29.

Fest, Joachim. *Hitler*. Translated by Richard Winston and Clara Winston. New York: Harcourt Brace Jovanovich, 1974.

Field, Geoffrey. *Evangelist of Race: The Germanic Vision of Houston Stewart Chamberlain*. New York: Columbia University Press, 1981.

Findlen, Paula. *Possessing Nature*. Berkeley: University of California Press, 1994.

Fleming, John. *The Philosophy of Zoology*. 2 vols. Edinburgh: Constable, 1822.

Fodor, Jerry. "Against Darwinism." *Mind and Language* 23 (2008): 1–24.

———. "Why Pigs Don't Fly." *London Review of Books* 29 (October 2007): 19–22.

Fodor, Jerry, and Massimo Piatelli-Palmarini. *What Darwin Got Wrong*. New York: Farrar, Straus and Giroux, 2010.

Forster, Michael. "Herder's Philosophy of Language, Interpretation, and Translation: Three Fundamental Principles." In *After Herder: Philosophy of Language in the German Tradition*. Oxford: Oxford University Press, 2010.

Gasman, Daniel. *Haeckel's Monism and the Birth of Fascist Ideology*. New York: Peter Lang, 1998.

———. *The Scientific Origins of National Socialism: Social Darwinism in Ernst Haeckel and the German Monist League.* New York: Science History Publications, 1971.

Ghiselin, Michael. "Darwin and Evolutionary Psychology." *Science* 179 (1973): 967.

———. "Darwin's Language Might Seem Teleological, but His Thinking Is Another Matter." *Biology and Philosophy* 9 (1994): 489–92.

Gieser, Suzanne. *The Innermost Kernel: Depth Psychology and Quantum Physics, Wolfgang Pauli's Dialogue with C. G. Jung*. New York: Springer, 2005.

Gliboff, Sander. *H. G. Bronn, Ernst Haeckel, and the Origins of German Darwinism*. Cambridge: MIT Press, 2008.

Gobineau, Joseph Arthur Grafen. *Versuch über die Ungleichheit der Menschenracen*. Translated by Ludwig Schemann. 2nd ed. 4 vols. Stuttgart: Fr. Frommanns Verlag, 1902–4.

Godfrey-Smith, Peter. "Explanation in Evolutionary Biology." *Mind and Language* 23 (2008): 32–41.

Golinski, Jan. *Making Natural Knowledge: Constructivism and the History of Science.* Cambridge: Cambridge University Press, 1998.

Gould, Stephen Jay. "*Abscheulich*! (Atrocious!) Haeckel's Distortions Did Not Help Darwin." *Natural History* 109, no. 2 (2000): 42–49.

———. "Eternal Metaphors of Palaeontology." In *Patterns of Evolution as Illustrated in the Fossil Record*, edited by A. Hallan. New York: Elsevier, 1977.

———. *Ever since Darwin*. New York: Norton, 1977.

———. *The Flamingo's Smile*. New York: Norton, 1985.

———. *The Hedgehog, the Fox, and the Magister's Pox*. New York: Harmony Books, 2003.

———. *Ontogeny and Phylogeny*. Cambridge: Harvard University Press, 1977.

———. *The Panda's Thumb*. New York: Norton, 1980.

———. *Wonderful Life: The Burgess Shale and the Nature of History*. New York: Norton, 1989.

Grant, Madison. *The Passing of the Great Race or the Racial Basis of European History*. New York: Charles Scribner's Sons, 1916.

———. *Der Untergang der großen Rasse* (The passing of the great race). Translated into German by Rudolf Polland. Munich: Lebmanns Verlag, 1925.

Gray, John. "The Atheist Delusion." *Guardian*, 15 March 2008, 4.

[Greg, William R.] "On the Failure of 'Natural Selection' in the Case of Man." *Fraser's Magazine* 78 (1868): 353–62.

Haeckel, Ernst. *Anthropogenie oder Entwickelungsgeschichte des Menschen*. Leipzig: Engelmann, 1874.

———. *Aus Insulinde: Malayische Reisebriefe*. Bonn: Emil Strauss, 1901.

———. *Generelle Morphologie der Organismen*. 2 vols. Berlin: Georg Reimer, 1866.

———. *Himmelhoch Jauchzend: Erinnerungen und Briefe der Liebe*. Edited by Heinrich Schmidt. Dresden: Carl Reissner, 1927.

———. *Kunstformen der Natur*. Leipzig: Bibliographisches Institut, 1904.

———. *Der Monismus als Band zwischen Religion und Wissenschaft*. Bonn: Emil Strauss, 1892.

———. *Die Natürliche Schöpfungsgeschichte*. Berlin: Reimer, 1868.

———. *Die Radiolarien. (Rhizopodia radiaria). Eine Monographie*. 2 vols. Berlin: Georg Reimer, 1862.

———. *Wanderbilder: Nach eigenen Aquarellen und Oelgemälden*. Gera-Untermhaus: W. Kochler, 1905.

———. *Die Weiträtsel, gemeinverständliche Studien über Monistische Philosophie*. Bonn: Emil Strauss, 1899.

———. *Zur Entwicklungsgeschichte der Siphonophoren*. Utrecht: C. van der Post, Jr., 1869.

Hale, Christopher. *Himmler's Crusade: The Nazi Expedition to Find the Origins of the Aryan Race*. New York: Wiley, 2003.

Hawkes, Nigel. "An Embryonic Liar." *Times* (London), 11 August 1997, 14.

Hawkins, Mike. *Social Darwinism in European and American Thought, 1860–1945*. Cambridge: Cambridge University Press, 1997.

Hecht, Günther. "Biologie und Nationalsozialismus." *Zeitschrift für die gesamte Naturwissenschaft* 3 (1937–38): 280–90.

Hegel, Georg Wilhelm Friedrich. *Vorlesungen über die Philosophie der Geschichte*. In *Werke*, vol. 12, edited by Eva Moldenhauer and Karl Michel. 4th ed. Frankfurt am Main: Suhrkamp, 1995.

Heisenberg, Werner. "Die Bewertung der modernen theoretischen Physik." *Zeitschrift für die gesamte Naturwissenschaft* 9 (1943): 201–12.

Herder, Johann Gottfried. *Abhandlung über den ursprung der Sprache*. In *Sprachphilosophische Schriften*, edited by Erich Heintel. Hamburg: Felix Meiner, 1975.

Hildebrandt, Kurt. "Positivismus und Natur." *Zeitschrift für die gesamte Naturwissenschaft* 1 (1935–36): 1–22.

———. "Die Bedeutung der Abstammungslehre für die Weltanschauung." *Zeitschrift für die gesamte Naturwissenschaft* 3 (1937–38): 15–34.

Hitler, Adolph. *Hitlers Zweites Buch*. Stuttgart: Deutsche Verlags-Anstalt, 1961.

———. *Mein Kampf*. Munich: Verlag Franz Eher Nachf., 1943.

———. *Monologe im Führer-Hauptquartier, 1941–1944*. Edited by Werner Jochmann. Munich: Albrecht Knaus, 1980.

Hoßfeld, Uwe. *Biologie und Politik: Die Herkunft des Menschen*. Erfurt: Landeszentrale für politische Bildung Thüringen, 2011.

———. *Geschichte der biologischen Anthropologie in Deutschland*. Stuttgart: Franz Steiner Verlag, 2005.

Hörbiger, Hanns, and Philipp Fauth. *Glazial-Kosmogonie*. Leipzig: R. Voigtländers Verlag, 1913.

Humboldt, Alexander von. *Cosmos*. 5 vols. Translated by E. C. Otté. New York: Harper and Brothers, 1848–68.

———. *Essai sur la géographie des plantes; accompagné d'un tableau physique des régions équinoxiales*. Paris: Chez Levrault, 1805.

———. *Kosmos. Entwurf einer physischen Weltbeschreibung*. 5 vols. Stuttgart: J. G. Cotta'scher Verlag, 1845–58.

Humboldt, Alexander von, and Aimé Bonpland. *Personal Narrative of Travels to the Equinoctial Regions of the New Continent, during the Years 1799–1804*. 7 vols. Translated by Helen Williams. London: Longman, Hurst, Rees, Orme, and Brown, 1818–29.

Humboldt, Wilhelm von. *Über die Kawi-Sprache auf der Insel Java*. 3 vols. Berlin: Königlichen Akademie der Wissenschaften, 1836.

Hume, David. *Treatise of Human Nature*. Edited by L. A. Selby-Bigge. Oxford: Clarendon Press, 1888. First published 1739.

Huxley, Thomas Henry. "Darwin on the Origin of Species." *Westminster Review*, n.s., 17 (1860): 541–70.

———. "Evolution and Ethics." In *Collected Essays.* 9 vols. New York: D. Appleton, 1896–1902. Essay originally published 1893.

———. "The Reception of the *Origin of Species.*" In vol. 1 of *Life and Letters of Thomas Henry Huxley*, edited by Leonard Huxley. 2 vols. New York: D. Appleton, 1900.

Jenkin, Fleeming. "The Origin of Species." *North British Review* 46 (1867): 277–318.

Jordan, Pascual. "Über den positivistischen Begriff der Wirklichkeit." *Die Naturwissenschaften* 22 (20 July 1934): 33–39.

Joyce, Richard. *The Evolution of Morality*. Cambridge: MIT Press, 2005.

Katz, Leonard, ed. *Evolutionary Origins of Morality*. Thorverton, UK: Imprint Academic, 2002.

Kershaw, Ian. *Hitler*. 2 vols. New York: Norton, 2000.

Kockerbeck, Christoph. *Ernst Haeckel's "Kunstformen der Natur" und ihr Einfluß auf die Deutsche Bildende Kunst der Jahrhundertwende*. Frankfurt: Peter Lang, 1986.

Kohn, David. "Darwin's Keystone: The Principle of Divergence." In *The Cambridge Companion to the "Origin of Species,"* edited by Michael Ruse and Robert J. Richards, 87–108. Cambridge: Cambridge University Press, 2009.

———. "Darwin's Principle of Divergence as Internal Dialogue." In *The Darwinian Heritage*, edited by David Kohn, 245–57. Princeton: Princeton University Press, 1985.

Krauße, Erika, ed. *Der Brief als Wissenschaftshistorische Quelle*. Berlin: Verlag für Wissenschaft und Bilding, 2005.

Krogh, Christian von. "Das 'Problem' Menschenwerdung." *Zeitschrift für die gesamte Naturwissenschaft* 6 (1940): 105–12.

Kuhn, Thomas. *The Structure of Scientific Revolutions*. 2nd ed. Chicago: University of Chicago Press, 1970.

Larson, Erik. *In the Garden of Beasts*. New York: Crown Books, 2011.

Lawn, Brian. *The Rise and Decline of the Scholastic "Quaestio Disputata."* Leiden: Brill, 1993.

Lee, Stephen. *Hitler and Nazi Germany*. London: Routledge, 2010.

Lehmann, Ernst. *Biologie im Leben der Gegenwart*. Munich: J. F. Lehmann Verlag, 1933.

Lennox, James. "Darwin *Was* a Teleologist." *Biology and Philosophy* 8 (1993): 409–21.

Levine, George. *Darwin the Writer*. New York: Oxford University Press, 2012.

Lewontin, R. C. "Adaptation." *Scientific American* 239 (1978): 212–28.

———. "Gene, Organism, and Environment." In *Evolution from Molecules to Men*, edited by D. S. Bendall, 273–85. Cambridge: Cambridge University Press, 1983.

———. "Organism and Environment." In *Learning, Development and Culture*, edited by E. C. Plotkin, 151–70. New York: Wiley, 1982.

———. *The Triple Helix: Gene, Organism and Environment*. Cambridge: Harvard University Press, 2000.

Lewontin, R.C., Steven Rose, and Leon Kamin. *Not in Our Genes*. New York: Pantheon, 1984.

Linnaeus, Carolus. *Systema naturae per regna tria naturae, secundum classes, ordines, genera, species, cum characteribus, differentiis, synonymis, locis*. 3 vols. Halle: Curt, 1760–70.

Locke, John. *An Essay concerning Human Understanding*. 2 vols. New York: Dover, 1959. First published 1670.

Lorenz, Konrad. "Nochmals: Systematik und Entwicklungsgedanken im Unterricht." *Der Biologe* 9 (1940): 24–36.

Lyell, Charles. *The Geological Evidences of the Antiquity of Man*. London: Murray, 1863.

———. *Principles of Geology*. 3 vols. Chicago: University of Chicago Press, 1987. First published in 1830–33.

MacRakis, Kristie. *Surviving the Swastika: Scientific Research in Nazi Germany*. Oxford: Oxford University Press, 1993.

Malthus, Thomas R. *An Essay on the Principle of Population*. 6th ed. 2 vols. London: Murray, 1826.

Marchant, James. *Alfred Russel Wallace: Letters and Reminiscences*. 2 vols. London: Cassell, 1916.

Marr, Wilhelm. *Der Sieg des Judenthums über das Germanenthum, vom nicht confessionellen Standpunkt aus betrachtet*. 8th ed. Bern: Rudolph Costenoble, 1879.

Mayr, Ernst. "Darwin's Principle of Divergence." *Journal of the History of Biology* 25 (1992): 343–59.

———. *The Growth of Biological Thought*. Cambridge: Harvard University Press, 1982.

McDonough, Frank. *Hitler and the Rise of the Nazi Party*. London: Pearson/Longman, 2003.

Miller, Kevin, and Ben Stein. *Expelled: No Intelligence Allowed*. Directed by Nathan Frankowski. Salt Lake City, UT: Rocky Mountain Pictures, 2008.

Milne-Edwards, Henri. *Introduction à la zoologie générale*. Paris: Victor Masson, 1851.

Moore, G. E. *Principia Ethica*. Cambridge: Cambridge University Press, 1903.

Müller, Friedrich Max. *Lectures on the Science of Language*. London: Longman, Green, Longman, and Roberts, 1861.

———. "The Science of Language." *Nature* 1 (1870): 256–59.

Newman, H. W. "Habits of Bombinatrices." *Transactions of the Entomological Society of London*, n.s., 1 (1850–51): 86–94.

*New York Times*. "A Little Riddle of the Universe." 27 July 1901.

Nordenskiöld, Erik. *The History of Biology*. Translated by L. B. Eyre. New York: Tudor, 1936. First published 1928.

Ospovat, Don. *The Development of Darwin's Theory: Natural History, Natural Theology, and Natural Selection, 1838–1859*. Cambridge: Cambridge University Press, 1981.

Paley, William. *Natural Theology*. London: Faulder, 1809.

———. *The Principles of Moral and Political Philosophy*. 2 vols. 16th ed. London: R. Faulder, 1806.

Parshall, Karen. "Varieties as Incipient Species: Darwin's Numerical Analysis." *Journal of the History of Biology* 15 (1982): 191–214.

Pearce, Trevor. "From 'Circumstances' to 'Environment': Herbert Spencer and the Origins of the Idea of Organism–Environment Interaction." *Studies in History and Philosophy of Biological and Biomedical Sciences* 41 (2010): 241–52.

———. " 'A Great Complication of Circumstances'—Darwin and the Economy of Nature." *Journal of the History of Biology* 43 (2010): 493–528.

Pennisi, Elizabeth. "Haeckel's Embryos: Fraud Rediscovered." *Science* 277 (1997): 1435.

Petrinovich, Lewis. *Human Evolution, Reproduction, and Morality*. New York: Plenum, 1995.

Popper, Karl. "Epistemology without a Knowing Subject." In *Objective Knowledge: An Evolutionary Approach*. Oxford: Oxford University Press, 1972.

*Quarterly Journal of Science*. Review of Darwin's *Origin of Species*. 3 (1866): 151–76.

Radick, Gregory. "Is the Theory of Natural Selection Independent of Its History?" In *The Cambridge Companion to Darwin*, 2nd ed., edited by Jonathan Hodge and Gregory Radick, 147–72. Cambridge: Cambridge University Press, 2009.

Ratner, Vadim. "Nikolay Vladimirovich Timoféeff-Ressovsky (1900–1981): Twin of the Century of Genetics." *Genetics* 158 (2001): 933–39.

Richards, Robert J. *Darwin and the Emergence of Evolutionary Theories of Mind and Behavior*. Chicago: University of Chicago Press, 1987.

———. "Darwin and Progress." *New York Review of Books* 52, no. 20 (15 December 2005).

———. "Darwin Tried and True." *American Scientist* 96 (May–June 2010): 238–42.

———. "A Defense of Evolutionary Ethics." *Biology and Philosophy* 1 (1986): 265–93.

———. "The Descent of Man: Review of *Darwin's Sacred Cause*." *American Scientist* 97 (September–October 2009): 415–17.

———. "The Epistemology of Historical Interpretation." In *Biology and Epistemology*, edited by Richard Creath and Jane Maienschein. Cambridge: Cambridge University Press, 2000.

———. "Ernst Haeckel's Alleged Anti-Semitism and Contributions to Nazi Biology." *Biological Theory* 2 (Winter 2007): 97–103.

———. *The Meaning of Evolution: The Morphological Construction and Ideological Reconstruction of Darwin's Theory*. Chicago: University of Chicago Press, 1992.

———. "The Moral Grammar of Narratives in History of Biology—The Case of Haeckel and Nazi Biology." In *The Cambridge Companion to the Philosophy of Biology*, edited by Michael Ruse and David Hull. Cambridge: Cambridge University Press, 2007.

———. "Neanderthals Need Not Apply." *New York Times Book Review*, 17 August 1997, 10.

———. "Race." In *The Oxford Companion to the History of Modern Science*, edited by John Heilbron. Oxford: Oxford University Press, 2001.

———. *The Romantic Conception of Life: Science and Philosophy in the Age of Goethe*. Chicago: University of Chicago Press, 2002.

———. *The Tragic Sense of Life: Ernst Haeckel and the Struggle over Evolutionary Thought*. Chicago: University of Chicago Press, 2008.

Richardson, Michael, J. Hanken, M. L. Gooneratne, C. Pieau, A. Raynaud, L. Selwood, and G. M. Wright. "There Is No Highly Conserved Embryonic State in the Vertebrates: Implications for Current Theories of Evolution and Development." *Anatomy and Embryology* 196 (1997): 91–106.

"Richtilinien für die Bestandsprüfung in den Volksbüchereien Sachsens." *Die Bücherei* 2 (1935): 279–80.

Rode, Karl. "Welt = Anschauung!" *Zeitschrift für die gesamte Naturwissenschaft* 2 (1936–37): 222–31.

Rokityanskij, Yakov. "N V Timofeeff-Ressovsky in Germany (July 1925–September 1945)." *Journal of Biosciences* 30, no. 5 (2005): 573–80.

Rosenberg, Alfred. *Der Mythus des 20. Jahrhunderts*. Munich: Hoheneichen Verlag, 1930.

Rütimeyer, Ludwig. "Review of *Natürliche Schöpfungsgeschichte* by Ernst Haeckel." *Archiv für Anthropologie* 3 (1868): 301–2.

Rupke, Nicolaas. "Neither Creation nor Evolution: The Third Way in Mid-Nineteenth Century Thinking about the Origin of Species." *Annals of the History and Philosophy of Biology* 10 (2005): 143–72.

Ruse, Michael. "Charles Darwin and Artificial Selection." *Journal of the History of Ideas* 36 (1975): 339–50.

———. *Darwinism and Its Discontents*. Cambridge: Cambridge University Press, 2008.

———. *Evolutionary Naturalism*. London: Routledge, 1995.

Ryback, Timothy. *Hitler's Private Library*. New York: Vintage Books, 2010.

Slack, J., P. Holland, and C. Graham. "The Zootype and the Phylotypic Stage." *Nature* 361 (1993): 490–92.

Schelling, Friedrich Wilhelm Joseph von. *Historisch-kritische Einleitung in die Philosophie der Mythologie* (1842). In *Ausgewählte Schriften*, edited by Manfred Frank. 6 vols. Frankfurt am Main: Suhrkamp, 1985.

Schlegel, August Wilhelm. *Observations sur la langue et la littérature provençales*. Paris: Librairie Grecque-Latine-Allemande, 1818.

Schleicher, August. *Die Darwinsche Theorie und die Sprachwissenschaft*. Weimar: Böhlau, 1863.

———. "Die Darwin'sche Theorie und die Thier- und Pflanzenzucht." *Zeitschrift für deutsche Landwirthe* 15 (1864): 1–11.

———. *Die Deutsche Sprache*. Stuttgart: Cotta'scher Verlag, 1860.

———. "Die ersten Spaltungen des indogermanischen Urvolkes." *Allgemeine Zeitschrift für Wissenschaft und Literatur* (August 1853): 786–87.

———. *Die Sprachen Europas in systematischer Übersicht*. Bonn: König, 1850.

———. *Über die Bedeutung der Sprache für die Naturgeschichte des Menschen*. Weimar: Böhlau, 1865.

———. *Zur vergleichenden Sprachengeschichte*. Bonn: H. B. König, 1848.

Schmidt, Johannes. "Schleicher." *Allgemeine deutsche Biographie* 31 (1890): 402–15.

Schweber, Silvan S. "Darwin and the Political Economists: Divergence of Character." *Journal of the History of Biology* 13 (1980): 195–289.

Sebright, John. *The Art of Improving the Breeds of Domestic Animals.* London: Howlett and Brimmer, 1809.

Secord, James. "Charles Darwin and the Breeding of Pigeons." *Isis* 72 (1981): 162–86.

Seidel, Eugen. "Die Persönalichkeit Schleichers." *Wissenschaftliche Beiträge der Friedrich-Schiller-Universität Jena* (1972): 8–17.

Shapin, Steven, and Simon Schaffer. *Leviathan and the Air-Pump*. Princeton: Princeton University Press, 1985.

Spencer, Herbert. *The Factors of Organic Evolution*. London: Williams and Norgate, 1887.

———. "Letter VII." *Nonconformist*, 19 October 1842.

———. "Mr. Martineau on Evolution." In *Recent Discussions in Science, Philosophy and Morals*. 2nd ed. New York: D. Appleton, 1882.

———. *Principles of Biology*. 2 vols. New York: D. Appleton, 1884. First published 1867.

———. *Social Statics: or, The Conditions Essential to Human Happiness Specified and the First of Them Developed*. London: Chapman, 1851.

[———]. "A Theory of Population, Deduced from the General Law of Animal Fertility." *Westminster and Foreign Quarterly Review* 57 (1852): 468–501.

Sober, Elliott. *Did Darwin Write the Origin Backwards? Philosophical Essays on Darwin's Theory*. Amherst, NY: Prometheus Books, 2011.

———. "Natural Selection, Causality, and Laws: What Fodor and Piatelli-Palmarini Got Wrong." *Philosophy of Science* 77 (2010): 594–607.

Stöcker, Adolf. *Das modern Judenthum in Deutschland besonders in Berlin*. Berlin: Verlag von Wiegandt und Grieben, 1880.

Stringer, Christopher, and Robin McKie. *African Exodus: The Origins of Modern Humanity*. New York: Holt, 1996.

Sulloway, Frank. "Darwin and His Finches: The Evolution of a Legend." *Journal of the History of Biology* 15 (1982): 1–53.

———. "Darwin's Conversion: The *Beagle* Voyage and Its Aftermath." *Journal of the History of Biology* 15 (1982): 327–98.

———. "Geological Isolation in Darwin's Thinking: The Vicissitudes of a Crucial Idea." *Studies in History of Biology* 3 (1979): 23–65.

Syllaba, Theodor. *August Schleicher und Böhmen*. Prague: Karls-Universität, 1995.

Tammone, William. "Competition, the Division of Labor, and Darwin's Principle of Divergence." *Journal of the History of Biology* 28 (1995): 109–31.

Taub, Liba. "Evolutionary Ideas and 'Empirical' Methods: The Analogy between Language and Species in Works by Lyell and Schleicher." *British Journal for the History of Science* 26 (1993): 171–93.

Theunissen, Bert. "Darwin and His Pigeons, the Analogy between Artificial Selection and Natural Selection Revisited." *Journal of the History of Biology* 45 (2012): 179–212.

Treitschke, Heinrich von. *Ein Wort über unser Judenthum*. Berlin: G. Reimer, 1880.

Trevor-Roper, Hugh. *The Last Days of Hitler*. 6th ed. Chicago: University of Chicago Press, 1992.

Wagner, Richard. *Das Judenthum in der Musik*. Leipzig: Weber, 1869.

Wallace, Alfred Russel. "A Defense of Modern Spiritualism." *Fortnightly Review,* n.s., 15 (1874): 630–57, 785–807.

———. "The Limits of Natural Selection as Applied to Man." In *Natural Selection and Tropical Nature*. London: Macmillan, 1891. Essay originally published 1870.

———. "The Origin of Human Races and the Antiquity of Man Deduced from the Theory of 'Natural Selection.'" *Anthropological Review* 2 (1864): clviii–clxxxvii.

———. "Review of *Principles of Geology* by Charles Lyell; *Elements of Geology* by Charles Lyell." *Quarterly Review* 126 (1869): 359–94.

Waters, C. Kenneth. "The Arguments in the *Origin of Species*." In *The Cambridge Companion to Darwin*, 2nd ed., edited by Jonathan Hodge and Gregory Radick, 120–43. Cambridge: Cambridge University Press, 2009.

Webb, Beatrice. *The Diary of Beatrice Webb: Volume One, 1873–1892*. Edited by Norman Mackenzie and Jeanne Mackenzie. Cambridge: Harvard University Press, 1982.

Wedgwood, Hensleigh. *On the Origin of Language*. London: Trübner, 1866.

Weikart, Richard. "Darwinism and Death: Devaluing Human Life in Germany, 1859–1920." *Journal of the History of Ideas* 63 (2002): 323–44.

———. *From Darwin to Hitler: Evolutionary Ethics, Eugenics, and Racism in Germany*. New York: Palgrave Macmillan, 2004.

———. *Hitler's Ethic: The Nazi Pursuit of Evolutionary Progress*. New York: Palgrave Macmillan, 2009.

———. "Was It Immoral for *Expelled* to Connect Darwinism and Nazi Racism?" 2 May 2008. Discovery Institute (http://www.discovery.org/a/5069).

Weindling, Paul. *Health, Race and German Politics between National Unification and Nazism, 1870–1945*. Cambridge: Cambridge University Press, 1989.

Welch, David. *Hitler*. London: Taylor and Francis, 1998.

Wells, Algernon. *On Animal Instinct*. Colchester: Longman, Rees, Orme, Brown, Green, and Longman, 1834.

Wells, Jonathan. *Icons of Evolution*. Washington, DC: Regnery, 2000.

———. *The Politically Incorrect Guide to Darwinism and Intelligent Design*. Washington, DC: Regnery, 2006.

Wessely, Christina. "Welteis. Die 'Astronomie des Unsichtbaren' um 1900." In *Pseudowissenschaft. Konzepte von Nicht/Wissenschaftlichkeit in der Wissenschaftsgeschichte*, edited by D. Rupnow, V. Lipphardt, J. Thiel, and C. Wessely. Frankfurt am Main: Suhrkamp, 2008.

Westenhöfer, Max. "Kritische Bemerkung zu neueren Arbeiten über die Menschenwerdung und Artbildung." *Zeitschrift für die gesamte Naturwissenschaft* 6 (1940): 41–62.

Whewell, William. *Astronomy and General Physics Considered with Reference to Natural Theology* (Bridgewater Treatise). Philadelphia: Carey, Lea and Blanchard, 1833.

Whitney, William Dwight. "Schleicher and the Physical Theory of Language." In vol. 1 of *Oriental and Linguistic Studies*, 298–337. 2 vols. New York: Charles Scribner's Sons, 1873.

Wilkinson, John. "Remarks on the Improvement of Cattle, etc. in a Letter to Sir John Sanders Sebright, Bart., M.P." MS. Nottingham, 1820.

Wilson, E. O. *Sociobiology*. Cambridge: Harvard University Press, 1975.

Wolpert, L. *The Triumph of the Embryo*. Oxford: Oxford University Press, 1991.

Wright, Chauncey. "Limits of Natural Selection." *North American Review* 111 (October 1870): 282–311.

Youatt, William. *Cattle: Their Breeds, Management, and Disease*. London: Library of Useful Knowledge, 1834.

*Zeitschrift für die gesamte Naturwissenschaft*. Editorial. 3 (1937–38): 1.

# INDEX